KT-145-611

£16.50

Replacement copy
Oct '79

DIFFUSION IN SOLIDS

McGraw-Hill Series in Materials Science and Engineering

AZÁROFF   *Introduction to Solids*

BARRETT AND MASSALSKI   *Structure of Metals*

ELLIOT   *Contribution of Binary Alloys, First Supplement*

PAUL AND WARSCHAUER   *Solids under Pressure*

SHEWMON   *Diffusion in Solids*

WERT AND THOMSON   *Physics of Solids*

DIFFUSION
IN SOLIDS

Paul G. Shewmon, *Department of Metallurgical Engineering,*
Carnegie Institute of Technology

McGraw-Hill Book Company
New York   San Francisco   Toronto   London

To

Robert Franklin Mehl

PREFACE

There are two main reasons for studying diffusion in solids. First, a knowledge of diffusion is basic to an understanding of the changes that occur in solids at high temperatures. Thus it is essential that the person interested in the kinetics of precipitation, oxidation, creep, annealing, etc., be acquainted with the fundamentals of diffusion. Also, it is often the case that a more detailed study of these processes requires the use of the more advanced concepts of diffusion in solids.

The second reason for studying diffusion is to learn more about how atoms move in solids. This is intimately connected with the study of point defects in solids and their movement. Point defects are the simplest type of defect in a solid, yet their concentration and movement cannot be observed directly. Since the number of such defects and their movement can be theoretically related to the diffusion coefficient, diffusion experiments have come to be the most frequently used means of studying point defects in solids.

The purpose of this book is to present a clear, concise, and relatively complete treatment of diffusion in solids. Throughout, the primary aim is to make clear the physical meaning and implications of the concepts which apply to diffusion in all crystalline solids. Where there was a choice of illustrative material, the author's background biased him toward choosing examples from metallic systems. However, those effects which are unique to nonmetallic systems are also treated in detail. Consistent with this emphasis on brevity and broad concepts, experimental results are quoted only in so far as they aid the reader in understanding the subject. Similarly, controversial issues have been avoided in favor of detailed discussion of what are felt to be the more firmly established fundamentals of the field.

The book has evolved from a one-semester course taught to first-year graduate students and should be suitable for use at the senior or graduate student level in a course on diffusion or in a course where diffusion is discussed as a basis for studying reactions in solids. The background required varies from chapter to chapter, but a student who has mastered sophomore engineering courses and a course in physical chemistry can understand the great majority of the book. A set of problems has been included at the end of each chapter as an additional means of helping the student come to grips with the material developed in the text and apply it to some related but novel situation.

The book should be helpful to the research worker who would like to learn more about the subject as a whole or wants a more detailed development of some topic than can be found in current papers on the subject. The chapters are largely independent of one another and the reader interested in only one topic will usually be able to fulfill his needs by reading all or parts of the chapter involved without reference to the preceding chapters.

Teaching from the notes of an embryonic text is an unsurpassed means of discovering the inadequacies of the presentation. The author is indebted to several classes of graduate students who have shown him the shortcomings of his efforts, both through their oral discussions and through their examination papers. Whatever clarity and value this book has a text stems primarily from these joint efforts. Many others have helped the author on particular points, but those individuals whose help was most frequent and most valuable were Professor G. T. Horne, whose judgment and patience were used repeatedly, and Professor C. Wert and Dr. L. Vassamillet, who carefully read the entire manuscript, pointing out errors and inadequacies.

P. G. Shewmon

CONTENTS

preface vii

Chapter 1. DIFFUSION EQUATIONS 1

    1-1. Fick's First Law 2
    1-2. Fick's Second Law 5
    1-3. Solutions to Fick's Law (Constant D) 6
    1-4. Kinetics of Precipitation 19
    1-5. Stress-assisted Diffusion 23
    1-6. Solutions for Variable D 28
    1-7. Diffusion in Noncubic Lattices 32

Chapter 2. ATOMIC THEORY OF DIFFUSION 40

    2-1. Random Movement and the Diffusion Coefficient . . 41
    2-2. Mechanisms of Diffusion 43
    2-3. Random-walk Problem 47
    2-4. Calculation of D 54
    2-5. Zener's Theory of D_0 62
    2-6. Empirical Rules for Obtaining ΔH and D_0 65
    2-7. Calculation of ΔH and ΔS from First Principles . . 67
    2-8. Experimental Determination of ΔH_v, ΔH_m, and ΔS_v . 71
    2-9. Divacancy Formation 78
    2-10. Effect of Hydrostatic Pressure on Diffusion 81

Chapter 3. DIFFUSION IN DILUTE ALLOYS 86

    3-1. Anelasticity Due to Diffusion 87
    3-2. Impurity Diffusion in Pure Metals 95
    3-3. Correlation Effects 100
    3-4. Diffusion in Dilute Binary Alloys 111

Chapter 4. DIFFUSION IN A CONCENTRATION
GRADIENT 115

    4-1. The Kirkendall Effect 116
    4-2. Darken's Analysis 117
    4-3. Phenomenological Equations 122
    4-4. Relationship between Chemical D_1 and Tracer D_1^* . . 125
    4-5. Test of Darken's Assumptions 127
    4-6. Ternary Alloys 130

ix

4-7. Diffusion in Multiphase Binary Systems. 132
4-8. Variation of $\tilde{D}$ across a Binary Phase Diagram . . . 133

Chapter 5. DIFFUSION IN NONMETALS. **137**

5-1. Defects in Ionic Solids 137
5-2. Diffusion and Ionic Conduction 140
5-3. Experimental Check of Relation between σ and D_T . 143
5-4. Effect of Impurities on D_T and σ 145
5-5. Effect of Impurities on Conductivity in Crystals with
      Frenkel Disorder 149
5-6. Relation of σ to D_T in AgBr (Frenkel Disorder) . . 151
5-7. Diffusion in Semiconductors 155
5-8. Diffusion in Ordered Alloys and Intermetallic Com-
      pounds 162

Chapter 6. HIGH-DIFFUSIVITY PATHS **164**

6-1. Analysis of Grain Boundary Diffusion 166
6-2. Experimental Results on Grain Boundary Diffusion . 171
6-3. Dislocation Effects 175
6-4. Diffusion Driven by Surface Tension. 179
6-5. Determination of D_s from Grain Boundary Grooving . 184
6-6. D_s from Field Emission Studies 186

Chapter 7. THERMAL DIFFUSION AND ELECTROLYSIS
            IN SOLIDS. **188**

7-1. Thermal Diffusion 189
7-2. Electrolysis of Solids 196

index **201**

chapter 1

DIFFUSION EQUATIONS

Changes in the structure of metals and their relation to physical and mechanical properties are the primary interest of the physical metallurgist. Since most changes in structure occur by diffusion, any real understanding of phase changes, homogenization, spheroidization, etc., must be based on a knowledge of diffusion. These kinetic processes can be treated by assuming that the metal is a continuum, that is, by ignoring the atomic structure of the solid. The problem then becomes one of obtaining and solving an appropriate differential equation. In this first chapter the basic differential equations for diffusion are given, along with their solutions for the simpler boundary conditions. The diffusion coefficient is also defined, and its experimental determination is discussed.

At no point in this chapter does the atomic nature of the material enter the problem. This is not meant to detract from the importance of the study of atomic mechanisms in diffusion, since the most interesting and most active areas of study in diffusion are, and will continue to be, concerned with the information that diffusion studies can contribute to the atomic models of solids. We initially omit models of diffusion to emphasize the types of problems that can be treated in this manner. In any theoretical development there are certain advantages and disadvantages to making the fewest possible assumptions. One advantage is that the results are quite generally applicable; a disadvantage is that the results are devoid of information about the atomic mechanism

1

of the process.   (Thermodynamics is an excellent example of this type of approach.)   The assumptions we make in Chap. 1 were first applied to the problem in 1855 by Adolf Fick.   It is indicative of the power of this approach that all the subsequent developments in the theory of solids have in no way affected the validity of the approach.

In the second and subsequent chapters we discuss the atomic proc-esses involved in diffusion.   In this chapter we present the basic differ-ential equations for diffusion and then develop several solutions, giving examples of the application of each.   The aim is to give the reader a feeling for the properties of the solutions to the diffusion equation and to acquaint him with those most frequently encountered.   Thus, no attempt is made at completeness or rigor.

1-1. FICK'S FIRST LAW

If an inhomogeneous single-phase alloy is annealed, matter will flow in a manner which will decrease the concentration gradients.   If the specimen is annealed long enough, it will become homogeneous and the net flow of matter will cease.   Given the problem of obtaining a flux equation for this kind of a system, it would be reasonable to take the flux across a given plane to be proportional to the concentration gradient across that plane.   For example, if the x axis is taken parallel to the concentration gradient of component 1 the flux of component 1 (J_1) along the gradient can be given by the equation

$$J_1 = -D_1 \left(\frac{\partial c_1}{\partial x} \right)_t \tag{1-1}$$

where D_1 is called the diffusion coefficient.   This equation is called Fick's first law and fits the empirical fact that the flux goes to zero as the specimen becomes homogeneous.   Although it need not have been the case, experiment shows that D_1, or equivalently the ratio of $-J_1$ to $\partial c_1/\partial x$, is independent of the magnitude of $\partial c_1/\partial x$.   In this respect Eq. (1-1) is similar to Ohm's law, where the resistance is independent of the voltage drop, or to the basic heat-flow equations in which the conductivity is independent of the magnitude of the temperature gradient.

To emphasize the dimensions of the terms, Eq. (1-1) is written again below with the dimensions of each term given in parentheses.

$$J \left(\frac{\text{mass}}{L^2 t} \right) = -D \left(\frac{L^2}{t} \right) \frac{\partial c}{\partial x} \left(\frac{\text{mass}/L^3}{L} \right)$$

The concentration can be given in a variety of units, but the flux must be put in consistent units. The diffusion coefficient is almost always given in terms of square centimeters per second. The units of concentration vary with the experiment and experimenter.

In a lattice with cubic symmetry, D has the same value in all directions, that is, the alloy is said to be isotropic in D. The assumption of isotropy will be made throughout the book unless a statement is made to the contrary. If there are other types of gradients in the system, other terms are added to the flux equation. These effects are interesting but complicated. They will be considered in Secs. 1-5 and 4-3.

As an example of the application of Eq. (1-1), consider the following experiment performed by Smith.[1] A hollow cylinder of iron is held in the isothermal part of a furnace. A carburizing gas is passed through the inside of the cylinder, and a decarburizing gas over the outside. When the carbon concentration at each point in the cylinder no longer changes with time, that is, $(\partial c/\partial t)_r = 0$, the quantity of carbon passing through the cylinder per unit time (q/t) is a constant. However, since J is the flow per unit area, it is a function of the radius r and is given by the equation

$$J = \frac{q}{At} = \frac{q}{2\pi r l t} \tag{1-2}$$

where l is the length of the cylinder through which carbon diffusion occurs. Combining Eqs. (1-1) and (1-2) gives an equation for q, the total amount of carbon which passed through the cylinder during the time t:

$$q = -D(2\pi l t)\frac{dc}{d\ln r} \tag{1-3}$$

For a given run, q, l, and t can be measured. If the carbon concentration through the cylinder wall is determined by chemical analysis, D can be determined from a plot of c versus $\ln r$. Such a plot will be a straight line if the diffusion coefficient does not vary with composition. However, for carbon in γ-iron Smith found that the slope of this plot $(dc/d\ln r)$ became smaller in passing from the low-carbon side of the tube to the high-carbon side. An example of his results for 1000°C is shown in Fig. 1-1. At this temperature the diffusion coefficient varies from 2.5×10^{-7} cm^2/sec at 0.15 weight per cent carbon to 7.7×10^{-7} cm^2/sec at 1.4 weight per cent carbon.

Similar experiments have frequently been performed by passing a gas through a metal membrane. Often the membranes are so thin it

[1] R. P. Smith, *Acta Met.*, **1**: 578 (1953).

is impossible to determine the concentration as a function of distance in the membrane by means of chemical analysis. The experimental results therefore consist of a measured steady-state flux, the pressure drop across the membrane, and the thickness of the membrane (Δy). This flux, for a given pressure drop, is called the permeability. To obtain a value of D from these data, the value of $\partial c/\partial y$ *inside* the membrane must be determined. One way to do this is to assume that

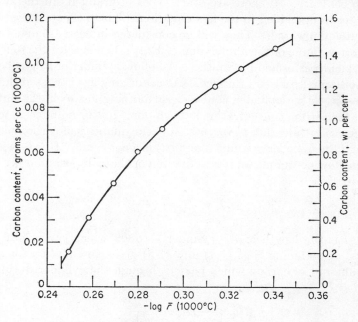

Fig. 1-1. $\bar{c}$ versus log $\bar{r}$ for a hollow cylinder of iron which attained a steady state with a carburizing gas passing through the inside and a decarburizing gas passing over the outside. [*From R. P. Smith, Acta Met.*, **1**: 578 (1953).]

the value of c in the metal at each gas-metal interface is the value that would exist in equilibrium with the gas if there were no net flux. This would be true if the solution of gas in the surface of the metal occurred much more rapidly than the diffusion out of the surface region into the rest of the metal. Experimentally, this assumption is checked by determining the fluxes for two thicknesses of membrane under the same pressure differential and at the same temperature. If equilibrium does exist at the gas-metal interface, then Δc is the same for both cases, and Eq. (1-1) gives

$$J = -D\frac{\Delta c}{\Delta y}$$

that is, J will be inversely proportional to Δy. At the other extreme, if the rate of solution of gas at the interface is what determines the flux, the flux will be the same for both values of membrane thickness, and no value of D can be obtained from the flux.

1-2. FICK'S SECOND LAW

If a steady state does not exist, that is, if the concentration at some point is changing with time, Eq. (1-1) is still valid, but it is not a convenient form to use. To obtain more useful equations, it is necessary to start with a second differential equation. It is obtained by using Eq. (1-1) and a material balance. Consider a bar of unit cross-sectional area with the x axis along its center. An element Δx thick along the x axis has flux J_1 in one side and J_2 out the other (see Fig. 1-2). If Δx is very small, J_1 can be accurately related to J_2 by the expansion

$$J_1 = J_2 - \Delta x \frac{\partial J}{\partial x} \qquad (1\text{-}4)$$

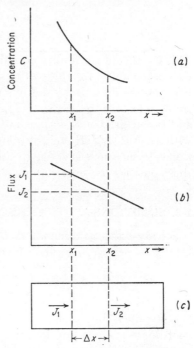

FIG. 1-2. (*a*) shows an assumed $c(x)$ plot, (*b*) shows $J(x)$ for this plot, and (*c*) shows the element of volume with the flux J_1 entering and J_2 leaving.

Since the amount of material that came into the element in unit time (J_1) is different from that which left (J_2), the concentration in the element has changed. The volume of the element is $1 \cdot \Delta x$ (unit area times the thickness), so the net increase in matter in the element can be expressed by any part of the equation

$$(J_1 - J_2) = \Delta x \frac{\partial c}{\partial t} = -\Delta x \frac{\partial J}{\partial x} \qquad (1\text{-}5)$$

Now Eq. (1-1) is valid at any instant even if the concentration and concentration gradient at that point are changing with time. Therefore we can substitute it into Eq. (1-5).

$$\frac{\partial c}{\partial t} = \frac{\partial}{\partial x}\left(D \frac{\partial c}{\partial x} \right) \qquad (1\text{-}6)$$

This is called Fick's second law of diffusion.

The next section will deal with Eq. (1-6); however, the generality of Eq. (1-5) should be pointed out. If one goes to three dimensions and uses a vector notation, the general statement of Eq. (1-5) is

$$\frac{\partial c}{\partial t} = -\nabla \mathbf{J} \tag{1-7}$$

This is called a continuity equation and stems only from the conservation of matter. In later sections we treat more complex cases in which Eq. (1-1) is no longer the flux equation. Nevertheless, we shall continue to be able to use Eq. (1-7) since it remains valid in the presence of additional gradients, e.g., a potential-energy gradient. If one deals with entities which are not conserved, e.g., vacancies, then an additional term which is equal to the rate of production or destruction of these entities per unit volume should be added to Eq. (1-7).

1-3. SOLUTIONS TO FICK'S SECOND LAW (CONSTANT D)

Steady-state Solutions. If D does not depend on position, and we take $\mathbf{J} = -D \nabla c$, Eq. (1-7) gives

$$\frac{\partial c}{\partial t} = D \nabla^2 c \tag{1-8}$$

$\nabla^2 c$ is called the Laplacian of c, and its representation in different coordinate systems can be found in many books dealing with applied mathematics.[1] If a steady state exists, $\partial c / \partial t = 0$, and the problem reduces to solving the equation

$$D \nabla^2 c = 0 \tag{1-9}$$

The simplest cases are, for cartesian coordinates,

$$D \frac{\partial^2 c}{\partial x^2} = 0 \tag{1-10}$$

For cylindrical coordinates,

$$D \left(\frac{\partial^2 c}{\partial r^2} + \frac{1}{r} \frac{\partial c}{\partial r} \right) = 0 \tag{1-11}$$

And for spherical coordinates,

$$D \left(\frac{\partial^2 c}{\partial r^2} + \frac{2}{r} \frac{\partial c}{\partial r} \right) = 0 \tag{1-12}$$

[1] J. Crank, "Mathematics of Diffusion," pp. 4–5, Oxford University Press, Fair Lawn, N.J., 1956.

The solutions to these equations are straightforward, and we shall not deal with them except as they arise in a few examples below.

The differential equation represented by Eq. (1-9) arises in many branches of physics and engineering. In two or more dimensions the solutions can be quite complicated. The interested reader should consult books on heat flow[1] or potential theory.[2]

Non-steady-state Solutions. If D is not a function of position, Eq. (1-6) becomes

$$\frac{\partial c}{\partial t} = D \frac{\partial^2 c}{\partial x^2} \tag{1-13}$$

We wish to determine the concentration as a function of position and time, that is, $c(x,t)$, for a few simple initial and boundary conditions. In general, the solutions of Eq. (1-13) for constant D fall into two forms. When the diffusion distance is short relative to the dimensions of the initial inhomogeneity, $c(x,t)$ can be most simply expressed in terms of error functions. When complete homogenization is approached, $c(x,t)$ can be represented by the first few terms of an infinite trigonometric series. (In the case of a cylinder, the trigonometric series is replaced by a series of Bessel functions.) We shall deal primarily with the former type and touch on the series solutions only in the case of precipitation. The reader who is interested in a comprehensive listing of solutions should consult Crank[3] or Carslaw and Jaeger.[1]

Thin-film Solution. Imagine that a quantity α of solute is plated as a thin film on one end of a long rod of solute-free material. If a similar solute-free rod is welded to the plated end of this rod (without any diffusion occurring) and the rod is then annealed for time t so that diffusion can occur, the concentration of solute along the bar will be given by the equation

$$c = \frac{\alpha}{2\sqrt{\pi D t}} \exp\left(-\frac{x^2}{4Dt}\right) \tag{1-14}$$

where x is the distance in either direction normal to the initial solute film. To show that Eq. (1-14) is the correct solution, two steps are necessary. First, differentiation shows that it is indeed a solution to

---

[1] H. S. Carslaw and J. C. Jaeger, "Conduction of Heat in Solids," Oxford University Press, Fair Lawn, N.J., 1959.

[2] O. D. Kellog, "Foundations of Potential Theory," Dover Publications, New York, 1953.

[3] J. Crank, "Mathematics of Diffusion," Oxford University Press, Fair Lawn, N.J., 1956.

Eq. (1-13).   Second, the equation satisfies the boundary conditions of the problem since

$$\text{for } |x| > 0 \qquad c \to 0 \text{ as } t \to 0$$
$$\text{for } x = 0 \qquad c \to \infty \text{ as } t \to 0$$

yet the total quantity of solute is fixed since

$$\int_{-\infty}^{\infty} c(x,t)\, dx = \alpha$$

The characteristics of this solution can best be seen with the help of Fig. 1-3.   Here the concentration is plotted against distance after some diffusion has occurred.   As more diffusion occurs, the $c(x)$ curve will spread out along the x axis. However, since the amount of solute is fixed, the area under the curve remains fixed.   To understand how this occurs, observe that $c(x = 0)$ decreases as $1/\sqrt{t}$ while the distance between the plane $x = 0$ and the plane at which c is $1/e$ times $c(x = 0)$ increases as $\sqrt{t}$.   This distance is given by the equation $x = 2\sqrt{Dt}$.

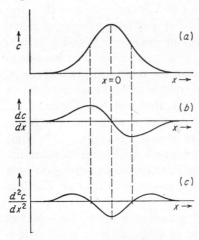

In Fig. 1-3b is plotted dc/dx versus x.   This is proportional to the flux across any plane of constant x. It will be seen that it goes to zero at $x = 0$ and at large positive or negative values of x.

Fig. 1-3. (a), (b), and (c) show $c(x)$, dc/dx versus x, and d^2c/dx^2 versus x, respectively, in arbitrary units for Eq. (1-14).

In Fig. 1-3c is plotted d^2c/dx^2 versus x.   This quantity is proportional to the rate of accumulation of solute in the region of any plane of constant x.   It is also proportional to the curvature of the $c(x)$ plot. Thus it is seen that in the region around $x = 0$, $c(x)$ is concave downward and the region is losing solute.   The concave upward regions on the outer portions of the $c(x)$ curve are gaining solute; regions at large values of x are undergoing no change in solute content.   To develop a feeling for these curves, the student is urged to derive the latter two for himself by plotting the slope of the curve above it versus x.

Equation (1-14) is often referred to as the solution for a thin film in the middle of an "infinite bar."   Since no bar is truly infinite, it is of value to consider just how long a bar must be for this equation to apply.

If the thin film is placed in the middle of a short bar and none of the solute is lost when it reaches an end of the bar, that solute which would normally have diffused past the end will be reflected back into the specimen, and c in that region will be higher than given by Eq. (1-14). Thus a short bar can be considered infinite if the quantity of solute which would lie outside its length in a truly infinite bar is an insignificant portion of the total solute present. Arbitrarily taking 0.1% as a sufficiently insignificant portion, we need to solve for x' in the equation

$$10^{-3} = \frac{\int_{x'}^{\infty} e^{-x^2/4Dt}\, dx}{\int_{0}^{\infty} e^{-x^2/4Dt}\, dx}$$

Here the numerator is proportional to the solute beyond x' in an infinite bar, and the denominator is proportional to the total solute in the bar. The numerator is called an error function and will be discussed more fully later.

The solution to this equation is $x' \simeq 4\sqrt{Dt}$. As might have been expected, the answer is stated in terms of the quantity $\sqrt{Dt}$. For sufficiently short times any bar is "infinite," and the time during which the bar can be considered infinite will depend on the magnitude of D as well as the elapsed time. The length $\sqrt{Dt}$ will appear in all diffusion problems, and the length of an infinite bar will be several times $\sqrt{Dt}$ in all cases.

Application of Thin-film Solution. Another property of Eq. (1-14) which is apparent from Fig. 1-3 is that at the plane $x = 0$, $dc/dx = 0$, and so the flux is zero. Thus if a thin layer of solute is applied to one end of a bar and allowed to diffuse in, the resulting solute distribution is described by Eq. (1-14) with $x \geq 0$ and $x = 0$ at the solute-rich end. After an appropriate anneal, thin sections are removed parallel to the initial interface. These are sections of constant x, and after the solute concentration of each is measured, a plot of $\ln c$ versus x^2 is made. From Eq. (1-14) it is seen that this is a straight line of slope $(4Dt)^{-1}$ so that if t is known, D can be calculated.

This procedure has been highly developed and is currently used for all the more accurate determinations of D for substitutional atoms. It is invariably used with a radioactive tracer as a solute since the concentration of a tracer can be determined with orders of magnitude greater sensitivity than is possible using chemical analysis. This means that D can be measured with such a small quantity of solute that the composition of the original sample is essentially unchanged. One of the other advantages of using radioactive tracers is that it is just as easy to study the diffusion of a silver tracer in silver as it is to study the diffusion of a cadmium tracer in silver. In both cases there is a concentration gradient for the tracer atom involved, and there will therefore be a spreading out of the tracer with time. The fact that the tracer is chemically *very* similar to the solvent in one case makes no difference in the application of the diffusion equations.

To help the reader develop an understanding of the magnitude of the values of D in metals and the procedures involved in their measurement, let us go through a rough calculation of the values of D which can be determined with this type of experiment. For the case of substitutional atoms in metals the value of D at the melting point is usually about 10^{-8} cm²/sec, so that this sets a rough upper limit on the values of D to be measured with this technique. If an accurate value of the slope of ln c versus x^2 is to be obtained, it is necessary to have several sections, and these sections should cover a concentration range of an order of magnitude. This means that sections should be taken to a depth of $x^2/4Dt \simeq 2.3$ or $x \simeq 3\sqrt{Dt}$. The minimum section depth which can be easily taken on a lathe is about 10^{-3} in. or 2.5×10^{-3} cm, so that for 10 sections the maximum value of x will be about 2.5×10^{-2} cm. If D is 10^{-8} cm²/sec, then it should be possible to make an accurate determination of D after an anneal of $t \simeq (2.5 \times 10^{-2}/3 \times 10^{-4})^2 \simeq 10^4$ sec $\simeq 2.5$ hr. It is normal practice to encapsulate the diffusion sample to protect it from oxidation. Under these conditions the specimen will probably not reach the furnace temperature for almost 10 min or 600 sec. What is actually measured experimentally is the Dt product, and since D changes rapidly with temperature, some correction must be made for this heating time. This can easily be done but not with great precision, so that it is preferable to work at diffusion times no shorter than a few hours even if other factors permit it.

The lower limit of the values of D that can be determined by this procedure can be estimated as follows. It is difficult to work with activities less than 0.002 of the value at $x = 0$. This requires that $x^2/4Dt \leq 5.5$. If the annealing time is one month, then $t \simeq 2.5 \times 10^6$ sec. Taking $x \simeq 2.5 \times 10^{-2}$ again gives $D \simeq 10^{-11}$ cm²/sec. By taking fewer sections or going to special sectioning techniques, it is possible to go down to $D = 10^{-13}$ or even 10^{-14} cm²/sec, but this can be done only with special techniques and in certain situations. It is seen then that D can be easily measured over three orders of magnitude using the sectioning technique. This may at first seem like an appreciable range, but because of the exponential variation of D with temperature, it allows the determination of D from the melting point down to only about three-fourths of the absolute melting temperature; there are many diffusion-controlled reactions which occur in solids at temperatures lower than this.

The reader particularly interested in the experimental determination of D by this technique should consult the review articles by Tomizuka[1] and Hoffman.[2]

Solution for a Pair of Semi-infinite Solids.   Consider the initial distribution which results if a piece of pure A is joined to pure B without interdiffusion.   This distribution is shown graphically in Fig. 1-4. The boundary conditions are given by

$$c = 0 \quad \text{for } x < 0, \text{ at } t = 0$$
$$c = c' \quad \text{for } x > 0, \text{ at } t = 0$$

A solution to the diffusion equation for this case can be obtained in the following manner: Imagine that the region of $x > 0$ consists of n slices, each of thickness $\Delta \alpha$ and unit cross-sectional area.   Consider one particular slice.   It initially contains $c' \Delta \alpha$ of solute, and if the surrounding regions were initially solute free, the distribution after some diffusion would be that given by the thin-film solution, i.e., Eq. (1-14).   The fact that there is solute in the adjacent slices does not in any way affect this result, and the actual solution is thus given by a superposition of the distributions from the individual slabs.   If α_i is the distance from the center of the ith slice to $x = 0$ (see Fig. 1-5), the concentration at any given value of x after time t will be

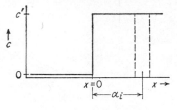

Fig. 1-4

$$c(x,t) \simeq \frac{c'}{2\sqrt{\pi Dt}} \sum_{i=1}^{n} \Delta \alpha_i \exp\left[-\frac{(x - \alpha_i)^2}{4Dt}\right] \qquad (1\text{-}15)$$

Figure 1-5 shows how these various exponentials superimpose to give the actual distribution for the case of rather thick slices.   In the limit of n going to infinity, $\Delta \alpha_i$ goes to zero, and from the definition of an integral

$$c(x,t) = \frac{c'}{2\sqrt{\pi Dt}} \int_0^\infty \exp\left[-\frac{(x - \alpha)^2}{4Dt}\right] d\alpha \qquad (1\text{-}16)$$

[1] C. Tomizuka, in K. Lark-Horowitz and V. A. Johnson (eds.), "Methods of Experimental Physics," vol. 6A, pp. 364–373, Academic Press, Inc., New York, 1959.

[2] R. E. Hoffman, in "Atom Movements," pp. 51–68, ASM, Cleveland, 1951.

Substituting $(x - \alpha)/2 \sqrt{Dt} = \eta$ we can rewrite the solution[1]

$$c(x,t) = \frac{c'}{\sqrt{\pi}} \int_{-\infty}^{x/2\sqrt{Dt}} \exp\left(-\eta^2\right) d\eta \tag{1-17}$$

This type of integral appears quite generally in the solutions of problems where the initial source of solute is an extended one and the diffusion distance $2\sqrt{Dt}$ is small relative to the length of the system. The integral cannot be evaluated in any simple manner, but because of its frequent appearance in diffusion and heat-flow problems, its

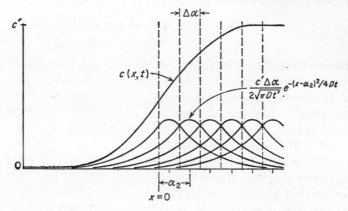

FIG. 1-5. $c(x,t)$ is the sum of the exponential curves which represent the solute diffusing out of each slab $\Delta\alpha$ thick.

values are available in tabular form.   The function given in Table 1-1 is called an error function and is defined by the equation

$$\text{erf}(z) = \frac{2}{\sqrt{\pi}} \int_0^z \exp\left(-\eta^2\right) d\eta \tag{1-18}$$

It can be shown that $\text{erf}(\infty) = 1$, and it is evident that

$$\text{erf}(-z) = -\text{erf}(z)$$

Equation (1-17) can thus be rewritten

$$c(x,t) = \frac{c'}{2}\left[1 + \text{erf}\left(\frac{x}{2\sqrt{Dt}}\right)\right] \tag{1-19}$$

This has already been plotted in Fig. 1-5.

[1] That this is a solution can be shown by differentiating Eq. (1-17) and substituting in Eq. (1-13).   The reader who is not familiar with differentiating an integral can find the procedure discussed in most books on advanced calculus or differential equations.

It should be noted that each value of the ratio c/c' is associated with a particular value of $z \equiv x/2 \sqrt{Dt}$. Thus $z = 1$ is always associated with $c/c' = 0.92$; the position of the plane whose composition is $0.92c'$ is given by the equation $x = 2 \sqrt{Dt}$. Further inspection shows that

TABLE 1-1. The Error Function

z	$\operatorname{erf}(z)$	z	$\operatorname{erf}(z)$
0	0	0.85	0.7707
0.025	0.0282	0.90	0.7970
0.05	0.0564	0.95	0.8209
0.10	0.1125	1.0	0.8427
0.15	0.1680	1.1	0.8802
0.20	0.2227	1.2	0.9103
0.25	0.2763	1.3	0.9340
0.30	0.3286	1.4	0.9523
0.35	0.3794	1.5	0.9661
0.40	0.4284	1.6	0.9763
0.45	0.4755	1.7	0.9838
0.50	0.5205	1.8	0.9891
0.55	0.5633	1.9	0.9928
0.60	0.6039	2.0	0.9953
0.65	0.6420	2.2	0.9981
0.70	0.6778	2.4	0.9993
0.75	0.7112	2.6	0.9998
0.80	0.7421	2.8	0.9999

SOURCE: The values of $\operatorname{erf}(z)$ to 15 places, in increments in z of 0.0001, can be found in the Mathematical Tables Project "Table of Probability Functions . . . ," vol. 1, Federal Works Agency, Works Projects Administration, New York, 1941. A discussion of the evaluation of $\operatorname{erf}(z)$, its derivatives and integrals, with a brief table is given by H. Carslaw and J. Jaeger, in Appendix II of "Conduction of Heat in Solids," Oxford University Press, Fair Lawn, N.J., 1959.

each composition moves away from the plane of $x = 0$ at a rate proportional to $\sqrt{Dt}$, with the exception of $c = c'/2$, which corresponds to $z = 0$ and thus remains at $x = 0$.

"Infinite" System—Surface Composition Constant. It has already been pointed out that the composition at the plane $x = 0$ in Eq. (1-19) does not change with time. Thus Eq. (1-19) in the region $x > 0$ can be used for the case in which an initially homogeneous alloy of solute c' is held in an atmosphere which reduces the surface concentration to

$c'/2$ and keeps it there for all $t > 0$. The boundary conditions are

$$c = \frac{c'}{2} \quad \text{for } x = 0, \text{ at } t > 0$$
$$c = c' \quad \text{for } x > 0, \text{ at } t = 0$$

and the solution is still Eq. (1-19). If the surface concentration is held at $c = 0$ instead of $c'/2$ for all $t > 0$, the solution becomes

$$c(x,t) = c' \operatorname{erf}\left(\frac{x}{2\sqrt{Dt}}\right) \tag{1-20}$$

If the surface concentration of an initially solute-free specimen is maintained at some composition c'' for all $t > 0$, solute is added to the specimen, and the solution is equivalent to Eq. (1-19) in the region $x < 0$. Since $\operatorname{erf}(-z) = -\operatorname{erf}(z)$, the solution for this case (in the region $x > 0$) is

$$c(x,t) = c'' \left[1 - \operatorname{erf}\left(\frac{x}{2\sqrt{Dt}}\right) \right] \tag{1-21}$$

Inspection of this equation shows that it fits the situation, since for $x = 0$, $c = c''$ and at $x \gg 2\sqrt{Dt}$, $c \simeq 0$.

It should be pointed out that in any of the solutions given by Eqs. (1-14) and (1-19) to (1-21), the zero of concentration can be shifted to fit the case where the initial or surface concentration is not zero, but some other constant value, say c_0. As an example, if the boundary conditions are

$$c = c_0 \quad \text{for } x > 0, \text{ at } t = 0$$
$$c = c' \quad \text{for } x < 0, \text{ at } t = 0$$

the solution is simply changed to

$$c(x,t) - c_0 = \frac{c' - c_0}{2} \left[1 - \operatorname{erf}\left(\frac{x}{2\sqrt{Dt}}\right) \right] \tag{1-22}$$

Application of Error-function Solutions.   The assumption of a constant D independent of position in the couple places a severe restriction on the use of this type of solution in making accurate determinations of D. If D is to be measured with a tracer in a chemically homogeneous alloy, it is usually easier to use the thin-film solution discussed earlier. On the other hand, if D is to be determined in a couple which has a range of chemical compositions in it, D will usually vary with position, i.e., composition, and the Matano-Boltzmann solution will be required (see Sec. 1-6 of this chapter).

In spite of these restrictions on the use of error-function solutions,

Eq. (1-19) has been used by Johnson[1] to accurately determine D for radioactive gold in a 50:50 Au-Ag alloy and to determine the chemical D in the range 45 to 55% Au.[2] His description of the procedure could not be improved on here, so the interested reader is referred to his paper.

The most frequent use of the error-function solutions arises when it is desired to estimate the amount of diffusion that will occur in a system where D is known to vary across the diffusion zone. A complete solution of the problem with a variable D is quite time-consuming, and essentially the same answer can be obtained by using an average value of D. This problem is found in the carburizing or decarburizing of steel, for which Eqs. (1-20) and (1-21) will often give adequate answers. Another case, in which these same equations would not give as accurate an answer, is in the dezincing of a Cu-Zn alloy. Here, as the zinc is removed, the sample shrinks, thus moving the original interface relative to the interior of the sample. No shrinkage was allowed for in the derivations of Eqs. (1-19) to (1-21); this further detracts from the accuracy of the answers obtained with these equations.

Separation of Variables. The above solutions have dealt with infinite systems. We now turn briefly to the type of solution which is simplest for "small" systems, that is, for those which approach complete homogenization. It is first assumed that there exist solutions which are the product of a function only of time $T(t)$ and a function of distance $X(x)$. That is, we assume that

$$c(x,t) = X(x) \, T(t) \tag{1-23}$$

It may be noted that the solutions discussed up to this point are excluded from this family since they are of the form $c(x,t) = f(x/\sqrt{t})$.

If we differentiate Eq. (1-23) in the prescribed manner and substitute in Fick's second law, the result is

$$X \frac{dT}{dt} = DT \frac{d^2X}{dx^2}$$

or

$$\frac{1}{DT} \frac{dT}{dt} = \frac{1}{X} \frac{d^2X}{dx^2} \tag{1-24}$$

---

[1] W. A. Johnson, *Trans. AIME*, **147**: 331 (1942).

[2] The chemical D is a measure of the rate of homogenization of an inhomogeneous alloy. This homogenization can occur by the movement of either or both components of the alloy. On the other hand, D for a particular tracer in a chemically homogeneous alloy measures the rate of movement of only that type of atom. The two D's measure different things so it should be no surprise that they differ in a given alloy. The relationship between the chemical D and the tracer D will be developed in Chap. 4.

The equation now contains only total differentials. The left side is a function only of time, and the right a function only of distance. But since x and t can be varied independently, Eq. (1-24) can be satisfied only if both sides of the equation are equal to a constant. This constant will be designated as $-\lambda^2$, where λ is a real number. The differential equation in time then is

$$\frac{1}{T}\frac{dT}{dt} = -\lambda^2 D$$

which integrates to

$$T = T_0 \exp\left(-\lambda^2 Dt\right) \tag{1-25}$$

where T_0 is a constant. The reason for requiring that the quantity $-\lambda^2$ have only negative values stems from our desire to deal only with systems in which any inhomogeneities disappear as time passes, i.e., that T approaches zero as t increases.

The equation in x is

$$\frac{d^2 X}{dx^2} + \lambda^2 X = 0$$

Since λ^2 is always positive, the solution to this equation is of the form

$$X(x) = A' \sin \lambda x + B' \cos \lambda x \tag{1-26}$$

where A' and B' are constants.

Combining Eqs. (1-25) and (1-26) gives

$$c(x,t) = (A \sin \lambda x + B \cos \lambda x) \exp\left(-\lambda^2 Dt\right)$$

But if this solution holds for any real value of λ, then a sum of solutions with different values of λ is also a solution. Thus in its most general form the product solution will be an infinite series of the form

$$c(x,t) = \sum_{n=1}^{\infty} (A_n \sin \lambda_n x + B_n \cos \lambda_n x) \exp\left(-\lambda_n^2 Dt\right) \tag{1-27}$$

Diffusion out of a Slab. As an example of the use of Eq. (1-27), consider the loss of material out both sides of a slab of thickness h. The boundary conditions to be assumed are

$$\begin{aligned} c &= c_0 \quad &&\text{for } 0 < x < h, \text{ at } t = 0 \\ c &= 0 \quad &&\text{for } x = h \text{ and } x = 0, \text{ at } t > 0 \end{aligned}$$

By setting all B_n equal to zero, c will be zero at $x = 0$ for all times. To make $c = 0$ at $x = h$, the argument of $\sin \lambda_n x$ must equal zero for

$x = h$. This is done by letting $\lambda_n = n\pi/h$, where n is any positive integer. If we substitute $B_n = 0$ and $\lambda_n = n\pi/h$ into Eq. (1-27), the first boundary condition requires that

$$c_0 = \sum_{n=1}^{\infty} A_n \sin \frac{xn\pi}{h} \tag{1-28}$$

To determine the A_n which will satisfy this equation, multiply both sides of this equation by $\sin (xp\pi/h)$, and integrate x over the range $0 \leq x \leq h$. This gives the equation

$$\int_0^h c_0 \sin \frac{xp\pi}{h} \, dx = \sum_{n=1}^{\infty} A_n \int_0^h \sin \frac{xp\pi}{h} \sin \frac{xn\pi}{h} \, dx$$

Each of the infinity of integrals on the right equals zero, except the one in which $n = p$. This integral is equal to $h/2$. The values of A_n which will satisfy Eq. (1-28) are thus given by the equation

$$A_n = \frac{2}{h} \int_0^h c_0 \sin \frac{n x \pi}{h} \, dx \tag{1-29}$$

Integration of this equation shows that $A_n = 0$ for all even values of n and $A_n = 4c_0/n\pi$ for odd values of n. Changing the summation index so that only odd values of n are summed over gives

$$A_n = A_j = \frac{4c_0}{(2j + 1)\pi} \qquad j = 0, 1, 2 \ldots \tag{1-30}$$

The solution is thus

$$c(x,t) = \frac{4c_0}{\pi} \sum_{j=0}^{\infty} \frac{1}{2j + 1} \sin \frac{(2j + 1)\pi x}{h} \exp \left[- \left(\frac{(2j + 1)\pi}{h} \right)^2 Dt \right]$$

$$\tag{1-31}$$

A moment's study of this equation shows that each successive term is smaller than the preceding one. Also, the percentage decrease between terms increases exponentially with time. Thus after a short time has elapsed, the infinite series can be satisfactorily represented by only a few terms, and for all time beyond some period t', $c(x,t)$ is given by a sine wave. To determine the error involved in using just the first term to represent $c(x,t)$ after some time t', it is easiest to consider the ratio of the maximum values of the first and second terms.

This ratio R is given by the equation

$$R = 3 \exp \frac{8\pi^2 Dt'}{h^2}$$

For $h = 2\sqrt{Dt}$, R is about 150, so that for $h^2 \leq 4Dt$ (or $t \geq h^2/4D$) the error in using the first term to represent $c(x,t)$ is less than 1% at all points.

This solution could be applied to the decarburization of a thin sheet of steel; and it is worthwhile to compare the use of this series solution with the error-function solution of Eq. (1-20). For short times the sheet thickness can be considered infinite. The carbon distribution below each surface will be given by the error-function solution as well as by this series solution. To evaluate $c(x,t)$ in this case using Eq. (1-31) would require the evaluation of many terms, and it is easier to look up the error function in a table. This is true until $h \simeq 3.2\sqrt{Dt}$, at which time $R \simeq 20$, and the error in using the error function is about 2% at the plane $x = h/2$. For times greater than this, the first term of Eq. (1-31) becomes a better approximation and would be used.

One of the most frequent metallurgical applications of this type of solution appears in the degassing of metals. Here it is often difficult to determine the concentration at various depths, and what is experimentally determined is the quantity of gas which has been given off or the quantity remaining in the metal. For this purpose the average concentration $\bar{c}$ is needed. This is obtained by integrating Eq. (1-31):

$$\bar{c}(t) = \frac{1}{h} \int_0^h c(x,t)\, dx = \frac{8c_0}{\pi^2} \sum_{j=0}^{\infty} \frac{1}{(2j+1)^2} \exp\left[-\left(\frac{(2j+1)\pi}{h}\right)^2 Dt \right]$$

$$(1\text{-}32)$$

The ratio of the first and second terms in this series is three times as large as in the case of Eq. (1-31), and for $\bar{c} \leq 0.8c_0$ the first term is an excellent approximation to the solution. The solution for $\bar{c}/c_0 \leq 0.8$ can be rewritten

$$\frac{\bar{c}}{c_0} = \frac{8}{\pi^2} \exp\left(-\frac{t}{\tau}\right) \qquad (1\text{-}33)$$

where $\tau \equiv h^2/\pi^2 D$ is called the relaxation time. Equation (1-33) is a type that is met frequently in systems that are relaxing to an equilibrium state. The quantity τ is a measure of how fast the system relaxes; when $t = \tau$, the system has traveled roughly two-thirds of the way from the initial state to the final state. Large values of τ thus characterize slow processes.

Equations similar to (1-32) and (1-33) and derived for the degassing of cylinders have been used in the accurate measurement of D. A good example is the study of the diffusion of hydrogen in nickel by Johnson and Hill.[1]

1-4. KINETICS OF PRECIPITATION

We consider next the solution of a much more complicated problem, namely, the kinetics of the removal of solute from a supersaturated matrix by the growth of randomly distributed precipitate particles. The average composition of the solute in solution, $\bar{c}(t)$, can be easily and continuously measured by several techniques. Thus the problem is to determine the relationship between the time variation of $\bar{c}(t)$ and the diffusion coefficient, particle shape, average interparticle distance, or other parameters that may be determined from independent observations or may need to be determined.

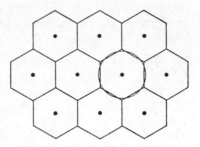

The problem is very complicated, but the complication comes primarily from the number of particles rather than the complexity of the diffusion around each particle. Thus the first task is to make a series of simplifying assumptions which make the problem tractable without making it so ideal-

FIG. 1-6. Close-packed plane out of close-packed lattice of precipitate particles, and traces of planes midway between particles. Circle is trace of equivalent sphere used in analysis of text.

ized that it bears no relation to the experimental system. A detailed mathematical analysis of the problem and the errors resulting from the different approximations has been given by Ham.[2]

Simplifications. Since the precipitate particles are randomly distributed, it is not unreasonable to approximate their distribution by that of a face-centered cubic space lattice. If planes are passed midway between all nearest-neighbor particles, these planes enclose each particle in a separate cell. [See Fig. 1-6 for a cut along the (111) plane.] Each of these planes is a plane of mirror symmetry. There is no net flux across a symmetry plane, since in the absence of a source or a sink in the symmetry plane, the only way it can be a symmetry

[1] M. Hill and E. Johnson, *Acta Met.*, **3**: 566 (1955).
[2] F. S. Ham, *J. Phys. Chem. Solids*, **6**: 335–351 (1958).

plane is for ∇c to be zero across it. Thus there will be no *net* flux into or out of each cell. Each cell can therefore be treated as if its walls were impermeable to solute. The problem of determining $\bar{c}(t)$ for the aggregate is thus reduced to determining $\bar{c}(t)$ for only one cell.

The mathematical description of the solution can be still further simplified without significant loss of accuracy if the cell bounded by segments of planes is replaced by a sphere of equivalent volume. The radius of this equivalent sphere is defined as r_e. We are now ready to proceed to the solution of the problem.

Solution for Short Times. We first consider the initial period of precipitation when the solute-drained region is small relative to the size of the equivalent sphere, that is, $r_e \gg \sqrt{Dt}$. The first equation is obtained from the fact that $\bar{c}$ decreases as solute atoms leave solution and precipitate on the precipitate surface. The amount of material leaving solution per unit time can be expressed both as the flux into the precipitate times the surface area of the precipitate and as the volume of the cell times the rate of change of $\bar{c}$ in the cell. If the precipitate is taken to be a sphere of radius α, the equality of these terms gives the equation

$$\frac{4\pi r_e{}^3}{3} \frac{d\bar{c}}{dt} = J(\alpha) 4\pi \alpha^2$$

where $J(\alpha)$ is the flux in the solid solution at $r = \alpha$.

To evaluate $J(\alpha)$ we assume that the actual solute distribution near $r = \alpha$ can be approximated by the steady-state solution that would satisfy the boundary conditions

$$\begin{aligned} c &= c_0 & \text{at } r = r_e \\ c &= c' & \text{at } r = \alpha(t) \end{aligned}$$

where c_0 is the initial concentration in the matrix and c' is the matrix concentration in equilibrium with the precipitate. Obviously this steady-state solution will not be obeyed throughout the equivalent sphere, but it is to be used only at $r = \alpha$ and then only for $\partial c/\partial r$ at $r = \alpha$. Letting α vary with time, the value of $\partial c/\partial r$ determined from this solution has the correct time dependence, and indeed for very dilute solutions the error involved in its use is insignificant.[1]

In spherical coordinates at the steady state, Fick's second law gives

$$\frac{\partial^2 c}{\partial r^2} + \frac{2}{r} \frac{\partial c}{\partial r} = 0$$

---

[1] F. C. Frank, *Proc. Roy. Soc.*, **A201**: 586 (1950).

The required solution is

$$c = \frac{b}{r} + d$$

Using the given boundary conditions

$$c = c_0 \quad \text{at } r = r_e$$
$$c = c' \quad \text{at } r = \alpha(t)$$

and assuming $r_e \gg \alpha(t)$ we have

$$c(r) = -\frac{(c_0 - c')\alpha}{r} + c_0$$

Thus

$$J(\alpha) = -D\left(\frac{\partial c}{\partial r}\right)_{r=\alpha(t)} = -D\frac{(c_0 - c')}{\alpha(t)}$$

or

$$\frac{d\bar{c}}{dt} = -\frac{3D}{r_e^3}(c_0 - c')\,\alpha(t) \tag{1-34}$$

A second equation can be obtained from the conservation of solute. Defining c_p as the concentration per unit volume of solute *in* the precipitate and assuming $\alpha = 0$ at $t = 0$, we have

$$\tfrac{4}{3}\pi c_p\, \alpha^3(t) = \tfrac{4}{3}\pi r_e^3[c_0 - \bar{c}(t)]$$

Solving this equation for $\alpha(t)$ gives

$$\alpha(t) = r_e\left[\frac{c_0 - \bar{c}(t)}{c_p}\right]^{\frac{1}{3}}$$

Substituting this in Eq. (1-34) gives the desired differential equation

$$\frac{d\bar{c}}{dt} = -\frac{3D(c_0 - c')}{c_p^{\frac{1}{3}}r_e^2}[c_0 - \bar{c}]^{\frac{1}{3}} \tag{1-35}$$

Defining $b \equiv 3D(c_0 - c')/c_p^{\frac{1}{3}}r_e^2$, Eq. (1-35) becomes

$$\frac{d\bar{c}}{dt} = -b(c_0 - \bar{c})^{\frac{1}{3}}$$

This integrates to

$$-\frac{3}{2}(c_0 - c)^{\frac{2}{3}} = -bt + \beta$$

but at $t = 0$, c_0 must equal $\bar{c}$, so $\beta = 0$.

Thus

$$\bar{c} = c_0 - \left(\frac{2bt}{3}\right)^{\frac{3}{2}} \tag{1-36}$$

This equation then gives the solution for short times. Ham shows that if $\alpha(0) = 0$, Eq. (1-36) is a good approximation for $\bar{c}/c_0 > \frac{2}{3}$. He also gives other solutions which are valid for more complete precipitation, that is, $\bar{c}/c_0 < \frac{2}{3}$, as well as a discussion of the case in which $\alpha(0) \neq 0$.

Let us now examine some of the properties of this solution. First, it is seen that for a spherical precipitate $c_0 - \bar{c}$ is proportional to $t^{\frac{3}{2}}$. This stems from the fact that the radius of the region drained of solute by the precipitate particle is proportional to $t^{\frac{1}{2}}$. Equation (1-36) simply states that $\bar{c}$ equals c_0 minus a term proportional to the quantity of solute precipitated. For spherical particles the volume drained of solute is proportional to $t^{\frac{3}{2}}$, and Eq. (1-36) results. If the precipitate particles were very long rods of fixed length, their volume or the volume of the region drained would change as r^2, and the last term in Eq. (1-36) would be replaced by a term in $(t^{\frac{1}{2}})^2$ or t. Finally, if the precipitate formed in sheets, e.g., all over a grain boundary, the volume of the drained region would increase as $t^{\frac{1}{2}}$, and the equation for $\bar{c}$ would change to the form $\bar{c} = c_0 - (\gamma t)^{\frac{1}{2}}$. We have not proved it, but equations of this form can be derived. Thus our model predicts that the precipitate shape can be determined from the initial slope of a plot of $\ln (\bar{c} - c_0)$ versus t.[1]

In the application of this analysis to experimental data, Eq. (1-36) is often replaced by an equivalent exponential equation. The function $\alpha \exp (-\beta x)$ can be expanded into the series

$$\alpha \exp (-\beta x) = \alpha \left(1 - \beta x + \frac{\beta^2 x^2}{2!} - \frac{\beta^3 x^3}{3!} \cdots \right) \tag{1-37}$$

which converges for all $\beta x < 1$. If $\beta x \ll 1$, the first terms $[\alpha(1 - \beta x)]$ give a good approximation. Comparing Eqs. (1-36) and (1-37), it is

---

[1] It should be remembered that the model used here is an approximation. Whether or not it is an acceptable approximation to any physically realizable case must be determined experimentally. Physically our model ignores the effects of surface tension and the effects of elastic strains set up by the phase being formed. Both of these can be important. It also ignores the problem of the stability of bumps formed on the particles by fluctuations in shape. To show the possible importance of this latter point, in a supercooled liquid an analysis similar to ours would dictate that the new phase would grow as spheres. However, owing to the fact that any small bump formed on the surface will grow faster than the sphere, the experimentally observed growth form is dendritic. The problem is obviously not simple. The justification for the treatment given here is not that it describes a situation that is proven to occur frequently in nature, but that it provides a well-developed example of how complicated problems can be treated. A minor benefit to the student is that this gives an inkling of how difficult it can be to obtain a model which is both physically adequate and mathematically tractable.

seen that for short times we can write

$$\bar{c} = c_0 \exp\left[-\left(\frac{2bt}{3c_0^{\frac{2}{3}}}\right)^{\frac{3}{2}}\right] \equiv c_0 \exp\left[-\left(\frac{t}{\tau}\right)^{\frac{3}{2}}\right] \qquad (1\text{-}38)$$

where the relaxation time τ is given by the expression

$$\tau = \frac{c_p^{\frac{1}{3}} r_e^2 c_0^{\frac{2}{3}}}{2D(c_0 - c')} \sim \frac{r_e^2}{2D}\left(\frac{c_p}{c_0}\right)^{\frac{1}{3}} \qquad (1\text{-}39)$$

Ham shows that for $\alpha(0) = 0$, this exponential equation fits the data down to smaller values of $\bar{c}/c_0$ than does Eq. (1-34). This agreement is because of compensating errors.

The relaxation time τ can be determined from data on $\bar{c}(t)$. Usually the quantities c_p, c_0, c', and D are known from other experiments. Thus r_e can be determined, and from this the mean interparticle distance can be calculated. The quantities that contribute to τ should be noted in Eq. (1-39). If τ increases, the rate of precipitation decreases; thus Eq. (1-39) indicates the variables which can be used to vary the rate of precipitation.

Solution for Longer Times. At longer times the concentration of solute at the cell boundary changes with time, and the best representation of the solution comes from the first term of a solution of the type

$$c(r,t) = c' + \sum A_n \exp\left(-\frac{t}{\tau_n}\right) f_n(r) \qquad (1\text{-}40)$$

This type of solution applies whether the precipitate forms as spheres, rods, or sheets, so the particle shape cannot be determined from $\bar{c}(t)$ in this time range. The added experimental difficulty of working at long times with slowly changing functions makes the solution of little practical interest.

1-5. STRESS-ASSISTED DIFFUSION

General Effect of Potential Gradient. As the last problem to be solved for constant D, we consider the effect of an elastic stress gradient on diffusion. This is representative of the problems in which Fick's first law is no longer the flux equation. A potential gradient tends to produce a flux of atoms, and this flux must be added to that produced by the concentration gradient to arrive at the equation for the total flux. In this section we consider the effect of a general potential gradient on the flux equation and the resulting changes in the equation for $\partial c/\partial t$.

Consider a single particle moving in a potential field $V(x,y,z)$; the potential gradient exerts a force **F** on the particle given by the equation

$$\mathbf{F} = -\nabla V \qquad (1\text{-}41)$$

As an example of this for macroscopic particles, consider a marble on an inclined plane. The potential here is due to gravity, and from elementary physics we know the force on the particle parallel to the plane to be proportional to the slope of the plane relative to the horizontal. Another common example of this type of force is the "pull" on a charged particle in an electrostatic potential gradient.

It is found empirically that a potential gradient or force gives rise to a mean diffusion velocity for the affected atoms. This fact is mathematically expressed by the equation

$$\mathbf{v} = B\mathbf{F} \qquad (1\text{-}42)$$

in which B is called the mobility and has the units velocity per unit force. It is worthy of note that this equation is not of the form "force equals mass times acceleration." The force gives rise to a steady-state velocity instead of a continuing acceleration because on the atomic scale atoms are continually changing their direction of motion and thus cannot accelerate under the action of a force in the way a free particle does. This intermittent motion of the atoms on an atomic scale will be discussed in detail in the next chapter. It is purposely avoided in this entire chapter to give the reader a clearer picture of the type of problems that can be treated with no assumptions concerning the atomic processes involved.

In applying a potential gradient instead of a concentration gradient, we are simply replacing one small force with another. Thus it is plausible, or even necessary, that the mobility is proportional to the diffusion coefficient D. In Sec. 4-3, we show that the relationship is

$$B = \frac{D}{kT} \qquad (1\text{-}43)$$

where k is Boltzmann's constant and T is the temperature in degrees Kelvin. The flux that results in a homogeneous system from **F** is thus the average velocity per particle times the number of particles per unit volume. If the units on **J** and c are consistent, we have

$$\mathbf{J} = c\mathbf{v} = B\mathbf{F}c = -\frac{Dc}{kT}\nabla V \qquad (1\text{-}44)$$

If a concentration gradient exists in addition to ∇V, the flux equation is given to a first approximation by the addition of Eqs. (1-1) and

(1-44) or

$$\mathbf{J} = -D\left(\nabla c + \frac{c\nabla V}{kT}\right) \tag{1-45}$$

Putting this flux equation in the continuity equation (1-7) gives, for constant D

$$\frac{\partial c}{\partial t} = D\nabla \cdot \left(\nabla c + \frac{c\nabla V}{kT}\right) \tag{1-46}$$

This then is the equation that needs to be solved to determine $c(x,y,z,t)$ in the presence of a potential gradient.

Solution for Very Short Times. The stress field around an interstitial atom in a solid solution is such that the atom can be attracted to a dislocation. Thus, in a supersaturated alloy, the precipitation rate on dislocations will be increased owing to the stress-induced drift which is superimposed on the drift due to any concentration gradient. If r is the radial distance between an interstitial atom and the core of an edge or screw dislocation, the interaction between the two can be approximated by the equations

$$V(r,\theta) = -\frac{\beta}{r} \quad \text{(screw)} \tag{1-47}$$

$$V(r,\theta) = -\frac{A}{r}\sin\theta \quad \text{(edge)} \tag{1-48}$$

where β and A are appropriately chosen constants.[1]

If an alloy is homogenized at a high temperature and quenched to a low temperature where it is supersaturated, the initial distribution for an isolated dislocation is given by

$$c = c_0 \quad \text{for } r > 0, \text{ at } t = 0$$

Expansion of Eq. (1-46) gives

$$\frac{\partial c}{\partial t} = D\nabla^2 c + \frac{D\nabla c \cdot \nabla V}{kT} + \frac{Dc\nabla^2 V}{kT} \tag{1-49}$$

Our aim is to determine the initial flux of atoms toward an isolated dislocation. This will depend on ∇c and ∇V. ∇V does not change with time, but ∇c does; and the general determination of how ∇c changes with time requires a solution of Eq. (1-49). However, for very short times the solution of this difficult problem can be avoided.

[1] A. Cochardt, G. Schoeck, and H. Wiedersich, *Acta Met.*, **3**: 533 (1955).

In the homogeneous alloy at $t = 0$, $\nabla c = 0$ and $\nabla^2 c = 0$, so the last term in Eq. (1-49) determines $\partial c / \partial t$. The equation for $\nabla^2 V(r,\theta)$ in cylindrical coordinates is

$$\nabla^2 V(r,\theta) = \frac{\partial^2 V}{\partial r^2} + \frac{1}{r}\frac{\partial V}{\partial r} + \frac{1}{r^2}\frac{\partial^2 V}{\partial \theta^2}$$

If $V(r,\theta)$ is given by Eq. (1-48), $\nabla^2 V = 0$, and $\partial c / \partial t = 0$. If $V = -\beta/r$, then $\nabla^2 V \neq 0$; but the resulting change in concentration with time sets up concentration gradients slowly, and for short times the drift of solute toward the dislocation can be satisfactorily approximated by assuming that $\nabla c = 0$.[1] Using Eq. (1-47) because of its relative simplicity, we obtain $\nabla V = \beta/r^2$. Using Eqs. (1-42) and (1-43), it is seen that this gradient moves the solute atoms toward the dislocations with a velocity given by the equation

$$v(r) = \frac{-dr}{dt} = \frac{D}{kT}\frac{\beta}{r^2} \tag{1-50}$$

Integrating between $r = r'$ at $t = 0$ and $r = 0$ at $t = t'$ gives

$$r' = \left(\frac{3D\beta t'}{kT}\right)^{\frac{1}{3}} \tag{1-51}$$

The interpretation of this equation is as follows. The atoms which were initially a distance r' from the core of the dislocation arrive at the dislocation core at $t = t'$, and other solute atoms which were initially at $r > r'$ have "taken their places" at $r = r'$. Thus, even though $\partial c / \partial t$ remains equal to zero, at $t = t'$ all the solute which was in the region $r < r'$ at $t = 0$ will have precipitated or segregated at the dislocation core.

The amount of solute q removed per unit length of dislocation after t' is given by the expression

$$q = c_0 \pi r'^2 = c_0 \pi \left(\frac{3D\beta t'}{kT}\right)^{\frac{2}{3}} \tag{1-52}$$

The period over which this solution is valid is determined by the volume in which the potential exerts an "appreciable" effect on the solute atoms. Let us be more precise. The thermal energy of a solute atom in the lattice will be about equal to kT. Thus when r becomes so large that $-V(r) < kT$, the potential energy will be less than the thermal energy of the particle, and the effect of the potential

---

[1] A more rigorous and more complete discussion of the mathematical analysis of stress-assisted precipitation has been given by F. S. Ham, *J. Appl. Phys.*, **30**: 915 (1959). A discussion of the approximations made herein is given there.

will be "inappreciable." We can thus define an "effective radius" for the potential as $r = R$, where

$$- V(R) = kT = \frac{\beta}{R} \tag{1-53}$$

Thus the condition $\partial c/\partial t \simeq 0$ will hold longer in the region $r < R$, where ∇V has an effect, than in the region $r \geq R$, where the effect of ∇V is insignificant. The solution embodied in Eq. (1-52) can apply only for the solute initially in the region $r' < R$. For times when $r' > R$, appreciable concentration gradients are set up, and a different analysis of the problem is required. The value of R for the case of carbon in α-Fe can be estimated from a value of β. Taking $\beta \simeq 10^{-20}$ dynes-cm²† gives $R \simeq 25$ A. With a dislocation density of $10^{11}/cm^2$, the mean distance between dislocations is about 300 A. Even with this relatively high dislocation density of $10^{11}/cm^2$, Eq. (1-52) breaks down after only a small percentage of the solute has precipitated. In an annealed material with a lower dislocation density, this simple drift approximation would apply for an even smaller fraction of the precipitation.

Solution for Intermediate Times. To study precipitation at longer times, we assume that the actual dislocation array can be replaced by a parallel set of dislocations whose spacing is the same as the mean spacing in the metal. Proceeding in a manner analogous to that used in Sec. 1-4, planes can be constructed midway between each dislocation pair and parallel to them. These are symmetry planes that completely enclose each dislocation in a cell. There will be no net flux into or out of each cell, so that each can be approximated by an isolated cylinder whose radius r_e is adjusted to give a cross-sectional area equal to that of the cell. The problem is then to study $\bar{c}(t)$ in one of these cylinders.

The treatment of this problem could be carried out in a manner similar to that for the spherical case of Sec. 1-4 if it were not for the potential around the dislocation. An equation for $\bar{c}$ during the time that $R > r'$ could easily be obtained from the preceding section. Precipitation for the period when $R < r'$ but $r_e \leq 2 \sqrt{Dt}$ is to be discussed here.

To discuss the fluxes inside the region $r < R$, it would be necessary to solve Eq. (1-49). However, we are not primarily interested in the solution $c(r,\theta,t)$, but in the rate at which solute is drained from the supersaturated solution. The solution for $c(r,\theta,t)$ can be avoided if it is observed that any atom entering the region $r < R$ will be drawn into

† Cochardt, Schoeck, and Wiedersich, *op. cit.*

the core at an ever-increasing rate.   That is, any solute that enters the region $r < R$ will be "captured" by the potential field and pulled irretrievably into the dislocation.   Thus once a solute atom enters the region $r < R$, it is removed from the solution just as if it had been precipitated.   The simplification thus becomes apparent.   The problem of solving for $c(r,t)$ when $\nabla V \neq 0$ can be sidestepped by specifying the boundary condition

$$c = c' \qquad r = R \qquad t > 0$$

The problem is then reduced to a simple diffusion problem with no potential and can be solved in a manner analogous to that used for the spherical precipitate.

In the above approximation, R is called the "capture radius" of the dislocation.   Its exact equivalence to R as defined by Eq. (1-53) has not been shown, but the existence of some capture radius at $r \leq R$ should be self-evident since $\nabla V(\sim 1/r^2)$ becomes very large at small r. A more exact definition of the capture radius is obtained by Ham from a solution of Eq. (1-49).   For the potential $V = -\beta/r$, his value of R is 1.78 times as large as the value given by Eq. (1-53).

Another problem in which the concept of a capture radius has been applied is in the kinetics of recombination of oppositely charged ions in germanium when the ions are initially homogeneously distributed. A complete description can be found in Reiss, Fuller, and Morin.[1]

1-6. SOLUTIONS FOR VARIABLE D

All of the solutions discussed so far have been valid only for constant D.   In real experiments the diffusion coefficient can, and will, vary. The diffusion coefficient for a given composition can vary with time, owing to changes in temperature.   It can also change with composition, and since there is a concentration gradient, this means that D changes with position along the sample.   In this latter case $D = D(x)$, and Fick's second law must be written

$$\frac{\partial c}{\partial t} = \frac{\partial}{\partial x}\left(D\,\frac{\partial c}{\partial x}\right) = \frac{\partial D}{\partial x}\,\frac{\partial c}{\partial x} + D\,\frac{\partial^2 c}{\partial x^2} \tag{1-54}$$

The term $\partial D/\partial x$ makes the equation inhomogeneous, and the solu-

---

[1] H. Reiss, C. Fuller, and F. Morin, *Bell System Tech. J.*, **35**: 535 (1956).   See also H. Reiss, *J. Appl. Phys.*, **30**: 1141 (1959), who gives a review of diffusion-controlled reactions in solids, with emphasis on inferring mechanisms from the results.

tion in closed form then ranges from difficult (for special cases) to impossible.

We will first discuss the solution for $D = D(c)$ which is most frequently used in solids and then show how to treat the case in which $D = D(t)$. For a more complete discussion of problems in which $D = D(c)$, the reader is referred to Crank.[1]

Boltzmann-Matano Analysis. This is the solution for $D = D(c)$ most commonly referred to in metallurgical diffusion studies. It will serve as an example of the different line of attack required. It does not give a solution $c(x,t)$ as obtained before, but allows $D(c)$ to be calculated from an experimental $c(x)$ plot. If the initial conditions can be described in terms of the one variable $\eta \equiv x/t^{\frac{1}{2}}$, c is a function only of η, and Eq. (1-54) can be transformed into an ordinary homogeneous differential equation. Using the definition of η, we have

$$\frac{\partial c}{\partial t} = \frac{dc}{d\eta}\frac{\partial \eta}{\partial t} = -\frac{1}{2}\frac{x}{t^{\frac{3}{2}}}\frac{dc}{d\eta}$$

and

$$\frac{\partial c}{\partial x} = \frac{dc}{d\eta}\frac{\partial \eta}{\partial x} = \frac{1}{t^{\frac{1}{2}}}\frac{dc}{d\eta}$$

Substituting in the first part of Eq. (1-54), we obtain

$$-\frac{x}{2t^{\frac{3}{2}}}\frac{dc}{d\eta} = \frac{\partial}{\partial x}\left(\frac{D}{t^{\frac{1}{2}}}\frac{dc}{d\eta}\right) = \frac{1}{t}\frac{d}{d\eta}\left(D\frac{dc}{d\eta}\right)$$

or

$$-\frac{\eta}{2}\frac{dc}{d\eta} = \frac{d}{d\eta}\left(D\frac{dc}{d\eta}\right) \tag{1-55}$$

This transformation of Eq. (1-54) into Eq. (1-55) is due to Boltzmann.[2] The method was first used to determine $D(c)$ by Matano.[3]

Consider the infinite diffusion couple which is described by the following initial conditions:

$$c = c_0 \quad \text{for } x < 0, \text{ at } t = 0$$
$$c = 0 \quad \text{for } x > 0, \text{ at } t = 0$$

Since $x = 0$ is excluded at $t = 0$ and the original concentration is not a function of distance aside from the discontinuity at $x = 0$, the initial conditions can be expressed in terms of η only as

$$c = c_0 \quad \text{at } \eta = -\infty$$
$$c = 0 \quad \text{at } \eta = \infty$$

[1] J. Crank, "Mathematics of Diffusion," chap. 9, Oxford University Press, Fair Lawn, N.J., 1956.

[2] L. Boltzmann, *Ann. Physik*, **53**: 960 (1894).

[3] C. Matano, *Japan. Phys.*, **8**: 109 (1933).

Since Eq. (1-55) contains only total differentials, we can "cancel" $1/d\eta$ from each side and integrate between $c = 0$ and $c = c'$, where c' is any concentration $0 < c' < c_0$

$$-\frac{1}{2}\int_{c=0}^{c=c'} \eta \, dc = \left[D \frac{dc}{d\eta} \right]_{c=0}^{c=c'}$$

The data on $c(x)$ are always at some fixed time so that substituting for η gives

$$-\frac{1}{2}\int_0^{c'} x \, dc = Dt \left[\frac{dc}{dx} \right]_{c=0}^{c=c'} = Dt \left(\frac{dc}{dx} \right)_{c=c'} \qquad (1\text{-}56)$$

The last equality in Eq. (1-56) comes from the fact that in this infinite system $dc/dx = 0$ at $c = 0$. From the additional fact that $dc/dx = 0$ at $c = c_0$, we have the condition

$$\int_0^{c_0} x \, dc = 0 \qquad (1\text{-}57)$$

so that Eq. (1-57) defines the plane at which $x = 0$. With this definition of x, $D(c')$ can be obtained from the graphical integration and differentiation of $c(x)$ using the equation

$$D(c') = -\frac{1}{2t} \left(\frac{dx}{dc} \right)_{c'} \int_0^{c'} x \, dc \qquad (1\text{-}58)$$

Application of Boltzmann-Matano Solution. The quantities needed to calculate a value of D are shown in Fig. 1-7. The Matano interface is the plane at which $x = 0$ in Eq. (1-57). Graphically, it is the line that makes the two hatched areas of Fig. 1-7 equal. The value of D at $c = 0.2c_0$ would be calculated by measuring the cross-hatched area of the figure and the reciprocal of the slope. The errors in the calculated values of $D(c)$ are largest at $c/c_0 \simeq 1$ and $c/c_0 \simeq 0$, since in these regions the integral is very small and dx/dc very large. In an effort to minimize these errors, the original concentration vs. distance data are usually plotted on probability paper, and the best line through the points used to make a plot similar to Fig. 1-7.[1]

This solution, which is quite useful for obtaining a composite value of D over a range of compositions, has been used during the last 25 years for what are now considered survey studies. For studies of the

---

[1] C. Wells, "Atom Movements," pp. 26–50, ASM, Cleveland, 1951.

mobilities of atoms in binary alloy, this technique has been largely superseded by tracer techniques.

Solutions for D a Function of Time. If D is a function of time but not of position, inspection of Eq. (1-54) shows that the equation reduces to $\partial c/\partial t = D(t)\,(\partial^2 c/\partial x^2)$. What this means is that all of the solutions which were used for constant D can be used, but the product

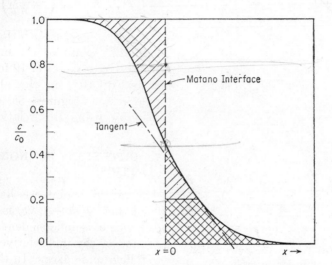

Fig. 1-7. The Matano interface is positioned to make the hatched areas on either side of it equal. The cross-hatched area and tangent show the quantities involved in calculating D at $c = 0.2c_0$.

Dt must be replaced by an averaged product designated $\overline{Dt}$ and given by the equation

$$\overline{Dt} = \int_0^t D(t)\,dt \qquad (1\text{-}59)$$

The most common application of this equation is to correct for the diffusion that occurs during the heating and cooling of a diffusion couple which has been annealed at some fixed temperature, although it can also be used to calculate the degree of homogenization achieved during a complicated annealing cycle.

As an example of the application of Eq. (1-59), consider a diffusion couple that has the temperature time history shown in Fig. 1-8a. The problem is to determine the time t' at temperature T' which would have produced the same amount of diffusion as actually occurred during the heating, annealing, and cooling. This can be determined graphically once the T versus t data are transformed into a plot of

D versus t. This has been done in Fig. 1-8b. It is seen that time spent in heating up to $0.8T'$ contributes nothing to the total amount of diffusion. This stems from the fact that $D(T)$ is given by an equation of the form

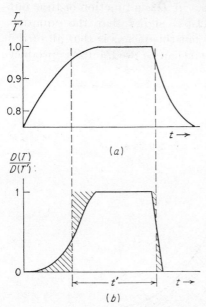

$$D = D_0 \exp\left(-\frac{Q}{RT}\right)$$

For many cases Q is such that near the melting point of the metal D increases by a factor of 10 for each increase of 10% in the absolute temperature. Figure 1-8b was obtained by using this relationship.

FIG. 1-8. (*a*) Temperature versus time record of diffusion sample; (*b*) D versus time for same sample. t' is the time at T' which would give the same amount of diffusion as actually occurred in the cycle.

1-7. DIFFUSION IN NONCUBIC LATTICES

Up to this point we have assumed that $D(\equiv -\mathbf{J}/\nabla c)$ is a constant which is independent of the direction of the flux relative to the crystallographic axes. In this section it will be shown that this assumption is true in any crystalline solid whose lattice has cubic symmetry. However, in noncubic crystals a complete description of the relation between the flux $\mathbf{J}$ and the concentration gradient ∇c requires the knowledge of two or more constants. As with the other topics discussed in this chapter, the proof of these properties of D requires no assumption about the atomic processes involved in diffusion. Rather, the results follow from the symmetry of the crystal and the properties of a second-order tensor.

The aim of the treatment given here is to develop an understanding of the physical basis of the results. The reasoning is rigorous but lacks elegance and generality. A general treatment requires a knowledge of the transformation properties of second-order tensors. The mathematics of the general treatment need not be difficult,[1] but if it

[1] W. Wooster, "A Text Book on Crystal Physics," chap. 1, Cambridge University Press, New York, 1949. A later, more general treatment has been given by J. Nye, "Physical Properties of Crystals," chap. 1, Clarendon Press, Fair Lawn, N.J., 1957.

is not, it is lengthy, and the physical aspects tend to get lost in the algebraic manipulations.

Consider the two vectors $\mathbf{J}$ and ∇c. In the most general case they will not be parallel, so that the simple relation $\mathbf{J} = -D\nabla c$, where D is a constant, is not adequate. For the general case it is assumed that a given component of the flux is influenced by each of the components of the gradient. Thus, the equations are[1]

$$J_x = -D_{11}\frac{dc}{dx} - D_{12}\frac{dc}{dy} - D_{13}\frac{dc}{dz} \qquad (1\text{-}60)$$

$$J_y = -D_{21}\frac{dc}{dx} - D_{22}\frac{dc}{dy} - D_{23}\frac{dc}{dz} \qquad (1\text{-}61)$$

$$J_z = -D_{31}\frac{dc}{dx} - D_{32}\frac{dc}{dy} - D_{33}\frac{dc}{dz} \qquad (1\text{-}62)$$

where the various scalar fluxes are parallel to the three axes of a cartesian coordinate system. The set of nine numbers designated D_{ij} is called a second-order tensor and is defined by the above equations.

To demonstrate the possible simplification of these equations, consider a cube of material whose lattice has cubic symmetry. Imagine that this cube is a single crystal with the cubic axes of the lattice perpendicular to the cube faces. A cartesian coordinate system is now set up with axes perpendicular to the cube faces (parallel to the cubic axes of the lattice), and a concentration gradient is established such that $dc/dx = \alpha \neq 0$ and $dc/dy = dc/dz = 0$. Equation (1-60) then gives the component of the flux parallel to the gradient as

$$J_x = -D_{11}\frac{dc}{dx}$$

If this gradient is removed and replaced by an equal gradient along the y axis, that is, $dc/dy = \alpha$ and $dc/dx = dc/dz = 0$, the component of the flux parallel to the gradient will be

$$J_y = -D_{22}\frac{dc}{dy}$$

Now if the lattice of the specimen has cubic symmetry, the x axis ([100] direction) and the y axis ([010] direction) are indistinguishable.

---

[1] A moment's reflection will show that these equations, and thus the treatment given here, can easily be adapted to the case of heat flow, the flow of electricity, or other cases in which a vector flux is related to a potential gradient. Thus, the results obtained below for D_{ij} hold equally well for the proportionality constants involved, e.g., the thermal and electrical conductivities.

Since the gradients were of equal magnitude in both cases, this symmetry requires that J_x in the first experiment must equal J_y in the second. But this says that $D_{11}\alpha = D_{22}\alpha$, which requires that $D_{11} = D_{22}$. Similarly the z axis ([001] direction) is indistinguishable from the x axis, so for a cubic crystal

$$D_{11} = D_{22} = D_{33} \tag{1-63}$$

In a lattice with tetragonal symmetry, the [010] and the [100] directions are identical but are distinguishable from the [001] direction. Thus in a tetragonal lattice

$$D_{11} = D_{22} \neq D_{33} \tag{1-64}$$

Finally, for an orthorhombic lattice each of the [100], [010], and [001] directions are distinguishable, so we have

$$D_{11} \neq D_{22} \neq D_{33} \tag{1-65}$$

For the case of hexagonal crystals, the above type of argument will prove that D_{11} is the same in each of the six close-packed directions of the basal plane. However, the fact that $D_{11} = D_{22}$, for the case of orthogonal axes cannot be proved without recourse to the transformation properties of second-order tensors. If it is accepted that $D_{11} = D_{22}$, then it should be apparent that for hexagonal crystals

$$D_{11} = D_{22} \neq D_{33}$$

To enlarge on these results, it is possible that in some hexagonal lattice D_{22} may be found experimentally to equal D_{33}. However, this cannot be asserted a priori from the symmetry. On the other hand, if it is reported that $D_{11} \neq D_{22}$ for some hexagonal material, then either the experimental results are incorrect or some type of defect, or field, was present which, in effect, destroyed the sixfold rotational symmetry about the [001] axis. The only other alternative would be to conclude that Eqs. (1-60) to (1-62) are inapplicable. This is conceivable but extremely improbable.

So far we have ignored the constants D_{ij} where $i \neq j$. However, if D_{ij} is to be a simple constant for any cubic crystal, then in addition to proving $D_{11} = D_{22} = D_{33}$ we must show that all $D_{ij} = 0$, if $i \neq j$. This can be done as follows. Again consider the single crystal cube with a cubic lattice. We establish a gradient such that $dc/dy = \alpha$, but $dc/dx = dc/dz = 0$. By measuring the component of the flux along the x axis, we can determine D_{12} since Eq. (1-60) gives

$$J_x = -D_{12}\frac{dc}{dy} = -D_{12}\alpha \tag{1-66}$$

Imagine next that the source and sink which established this gradient are removed and the crystal is rotated 180° about its x axis ([100] direction), relative to the source and sink. Since the [100] direction has fourfold rotational symmetry, the original and final positions of the lattice will be indistinguishable. If the source and sink are again applied, the gradient is again established along the y axis. However, since the y axis is fixed in the crystal, the 180° rotation has interchanged the $+y$ and the $-y$ axes, so the gradient will be just the negative of its previous value. This means that now $dc/dy = -\alpha$, and $dc/dx = dc/dz = 0$. This gives

$$J_x = -D_{12} \frac{dc}{dy} = D_{12}\alpha \tag{1-67}$$

But by symmetry, J_x from Eq. (1-66) must equal J_x from Eq. (1-67), or

$$D_{12}\alpha = -D_{12}\alpha$$

This requirement of symmetry can be satisfied only if $D_{12} = 0$. If the gradient had been along the z axis, the same rotation could have been used to show that $D_{13} = 0$. Since the y and the z axes also have two-fold rotational symmetry, it follows that all the off-diagonal terms (that is, D_{ij}, $i \neq j$) are zero. Finally, we have used only twofold rotational symmetry, so from the same proof it follows that the D_{ij} also equal zero when $i \neq j$ for tetragonal and orthorombic crystals.

Similar reasoning shows that the $D_{ij}(i \neq j)$ terms are zero for lattices which have hexagonal symmetry. The results can be summarized as follows:

Cubic $\qquad\qquad D_{ij} = \begin{bmatrix} D_{11} & 0 & 0 \\ 0 & D_{11} & 0 \\ 0 & 0 & D_{11} \end{bmatrix}$ $\qquad$ (1-68)

Tetragonal
and hexagonal $\qquad D_{ij} = \begin{bmatrix} D_{11} & 0 & 0 \\ 0 & D_{11} & 0 \\ 0 & 0 & D_{33} \end{bmatrix}$ $\qquad$ (1-69)

Orthorhombic $\qquad D_{ij} = \begin{bmatrix} D_{11} & 0 & 0 \\ 0 & D_{22} & 0 \\ 0 & 0 & D_{33} \end{bmatrix}$ $\qquad$ (1-70)

To demonstrate the use of these conclusions and Eqs. (1-60) to (1-62), we shall apply them to the case where D_{ij} is given by Eq. (1-69). Consider the determination of the diffusion coefficient in such a material using a tracer. If a single crystal is available with the c axis (the sixfold or fourfold axis) normal to one face, the tracer can be deposited on that face. The concentration gradient is parallel to the c axis,

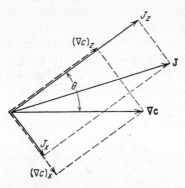

FIG. 1-9. The orientation of the flux **J** and the concentration gradient **∇c** for the case $D_{11} < D_{33}$. The z axis is parallel to the c axis of the lattice. **∇c** is normal to the y axis so that $(\nabla c)_y$ and J_y equal zero.

since $D_{13} = D_{23} = 0$, and so is the flux. Thus if $c(u)$ is determined by taking sections parallel to the initial face, one obtains D_{33} from a plot of $\ln c$ versus u^2. If the c axis is in the face covered with tracer, the flux and gradient are again parallel, and the determination of D_{11} is straightforward.

For the intermediate case where the plated face makes an angle of $90° - \theta$ with the c axis, the concentration gradient makes an angle of θ with the c axis. In this case the flux and the gradient are not parallel if $D_{11} \neq D_{33}$. This is shown in Fig. 1-9, where $D_{33} > D_{11}$. If $c(u)$ is now determined, where u is the distance normal to the plated face, the value of D obtained from a plot of $\ln c$ versus u^2 is called $D(\theta)$ and is given by the equation[1]

$$D(\theta) = D_{33} \cos^2 \theta + D_{11} \sin^2 \theta \qquad (1\text{-}71)$$

If necessary, D_{33} and D_{11} can be determined from measurements of $D(\theta)$ for different values of θ.

Diffusion coefficients have been carefully measured in only a few noncubic materials. Much of the available data is summarized in

TABLE 1-2. $D_{\parallel}$ and $D_{\perp}$ for Several Metals

Metal	Ref.	Cryst. struct.	$D_{0\perp}\left(\dfrac{cm^2}{sec}\right)$	$D_{0\parallel}\left(\dfrac{cm^2}{sec}\right)$	$\Delta H_{\perp}\left(\dfrac{kcal}{mole}\right)$	$\Delta H_{\parallel}\left(\dfrac{kcal}{mole}\right)$	$(D_{\parallel}/D_{\perp})$ at $T=0.9T_m$
Cd	1	hcp	0.1	0.05	19.1	18.2	1.20
Mg	2	hcp	1.5	1.0	32.5	32.2	0.82
Tl	3	hcp	0.4	0.4	22.6	22.9	0.73
Sn	4	bct	1.4	8.2	23.3	25.6	0.48
Zn	5	hcp	0.58	0.13	24.3	21.8	1.67

[1] E. Wajda, G. Shirn, and H. Huntington, *Acta Met.*, **3**: 39 (1955).
[2] P. Shewmon, *Trans. AIME.* **206**: 918 (1956).
[3] G. Shirn, *Acta Met.*, **3**: 87 (1955).
[4] J. Meakin and E. Klokholm, *Trans. AIME*, **218**: 463 (1960).
[5] G. Shirn, E. Wajda, and H. Huntington, *Acta Met.*, **1**: 513 (1953).

[1] Nye, *op. cit.*, chap. 1, pp. 24–26.

Table 1-2. Here D_{11} and $D_\perp$ refer to the diffusion coefficients parallel and perpendicular to the c axis of the lattice.

PROBLEMS

1-1. An experiment similar to Smith's (Sec. 1) is performed on a hollow iron tube with a 1.11-cm outer diameter, a 0.86-cm inner diameter, and a length of 10 cm. In 100 hr, 3.60 g of carbon passes through the tube. The variation of carbon with radius in the tube is given below. Plot $\bar{c}$ versus $\bar{r}$, and from this calculate and plot the diffusion coefficient over this range of carbon contents.

$\bar{r}$ (cm)	Wt. % Carbon	$\bar{r}$ (cm)	Wt. % Carbon
0.553	0.28	0.491	1.09
0.540	0.46	0.479	1.20
0.527	0.65	0.466	1.32
0.516	0.82	0.449	1.42
0.503			

1-2. It is desired to store hydrogen at 400°C. A steel tank is available, and you are asked to calculate the steady-state rate of pressure drop for a cubical tank of 1-l capacity with a wall thickness of 1 mm when the internal pressure is 132 psi absolute. The tank is to operate in a vacuum. Assume that $D = 10^{-4}$ cm²/sec, independent of the state of stress, that the hydrogen in the steel is in equilibrium with the atmosphere at both the high- and low-pressure sides, that the solubility is proportional to $\sqrt{P}$, and that at $P = 1$ atm the solubility is 3 ppm (by weight). State clearly any additional assumptions made in your analysis.

1-3. Derive the expression relating the thermal diffusivity d (cm²/sec) in the expression $\partial T/\partial t = d\,\partial^2 T/\partial x^2$ to the thermal conductivity k (cal/cm sec °C) the specific heat C_p (cal/g °C), and the density ρ (g/cm³)

(a) Using a heat balance on an element of volume similar to the material balance used in the text to derive Fick's second law.

(b) Using dimensional analysis.

1-4. By differentiation and substitution, show that the solutions to Eqs. (1-11) and (1-12) are

$$c(r) = \alpha_1 \ln r + \beta_1$$

and

$$c(r) = \frac{\alpha_2}{r} + \beta_2$$

1-5. A sphere of inner radius r_i and outer radius r_0 is immersed in a bath of temperature T_0. If the inner temperature is kept at T_i and heat is lost at a rate $\dot{q}$, derive an equation for the thermal diffusivity d in terms of r_i, r_0, T_i, T_0, and $\dot{q}$. d is defined by the equation.

$$\frac{\partial T}{\partial t} = d\left(\frac{\partial^2 T}{\partial r^2} + \frac{2}{r}\frac{\partial T}{\partial r}\right)$$

1-6. Prove that the net flux across a symmetry plane is zero in a three-dimensional medium where D is a constant. [Hint: A symmetry plane is said to exist on the plane $x = 0$ if $c(x,y,z) = c(-x,y,z)$.]

1-7. Show by differentiation and substitution that

$$c(x,t) = \frac{\alpha}{\sqrt{t}} \int_{-\infty}^{\infty} f(x') \exp\left[\frac{-(x-x')^2}{4Dt}\right] dx'$$

is a solution to the diffusion equation

$$\frac{\partial c}{\partial t} = \frac{D \, \partial^2 c}{\partial x^2}$$

[If $f(x')$ is the initial distribution of solute for an infinite system, this integral gives $c(x,t)$ for $t > 0$.]

1-8. A piece of 0.1% C steel is to be carburized at 930°C until the carbon content is raised to 0.45% C at a depth of 0.020 in. (0.050 cm). The carburizing gas holds the surface concentration at 1% carbon for all $t > 0$. If $D \simeq 1.4 \times 10^{-7}$ cm²/sec for all compositions,

(a) Calculate the time required at the carburizing temperature.

(b) What time is required at the same temperature to double this amount of penetration?

(c) If $D = 0.25 \exp(-34,500/RT)$ cm²/sec, what temperature is required to get 0.45% C at 0.040 in. (0.10 cm) in the same time as 0.02 in. was attained at 930°C?

1-9. Given an iron plate 7 in. thick, how long will it take to remove 90% of the hydrogen during an anneal in which the surface concentration of hydrogen in the iron is held at zero, if D for hydrogen is 10^{-4} cm²/sec?

1-10. It has been suggested that the change in hydrogen potential of an aqueous solution could be measured by the change in electrical resistance of an inert metal foil immersed in the solution. (The resistance of the foil is proportional to the average concentration of hydrogen in the metal.) If the diffusion coefficient of hydrogen in the metal is 10^{-5} cm²/sec, what should the thickness of the foil be if a resistance measurement is to indicate within 1 sec the magnitude of a sudden change in the hydrogen potential to within 5%?

1-11. A bar of pure copper was joined to a bar of a copper–5% nickel alloy; after annealing for 10^5 sec, sections are taken parallel to the plane of joining with a lathe. Arbitrarily taking the center of the first cut in which nickel was found to be zero on the x axis, the composition varied as follows:

% Ni	$\bar{x}$ (10^{-3} in.)	% Ni	$\bar{x}$ (10^{-3} in.)
0.10%	0	3.05	16
0.21	2	3.69	18
0.41	4	4.02	20
0.67	6	4.32	22
1.05	8	4.63	24
1.31	10	4.82	26
1.96	12	4.92	28
2.52	14		

(a) What is the diffusion coefficient, assuming D to be independent of composition?

(b) Does D vary with composition?

1-12. In a pure gold wire quenched from 700°C, it is thought that the supersaturation of vacancies is relieved by adsorption of vacancies at dislocation lines.

Considering the dislocation lines to be fixed cylindrical sinks of constant radius r_0, derive an equation giving the time dependence of the ratio of the average vacancy concentration $\bar{c}(t)$ to the initial concentration c_0 (for $0.8 < \bar{c}/c_0 < 1$), which could be used to check this hypothesis.

1-13. In a quenched specimen similar to that described above, derive the equation for the initial change of $\bar{c}/c_0$ with time for the case in which only grain boundaries act as sinks for the vacancies.

1-14. Show that for those initial conditions in which the Boltzmann substitution $c(x,t) = c(\eta)$ is permissible, the distance between any two compositions increases as $t^{\frac{1}{2}}$, even when D varies with composition.

1-15. A rod of pure copper was joined to a rod of a 29.4% Zn–70.6% Cu alloy. After annealing for 360 hr the % Zn vs. distance data were plotted on probability paper, and the following values were picked off from the best line through the data:

Atomic % Zn	$\bar{x}$ (10^{-2} cm)	Atomic % Zn	$\bar{x}$ (10^{-2} cm)
0.3	50.05	23.5	36.55
1.5	48.15	25.0	34.05
4.4	46.45	26.5	30.75
8.8	44.95	27.9	25.15
14.7	43.15	28.8	18.95
20.6	39.65	29.1	14.95

Determine the position of the Matano interface, and calculate $D(c)$ at 5% Zn intervals across the couple.

1-16. A thin film of radioactive copper was electroplated on the end of a copper cylinder. After a high temperature anneal of 20 hr, the specimen was sectioned, and the activity of each section counted. The data are

a (counts/min/mg)	$\bar{x}$ (10^{-2} cm)
5012	1
3981	2
2512	3
1413	4
524.8	5

(a) Plot the data and determine D.

(b) Calculate the slope of log a versus $\bar{x}^2$ using a least-squares procedure, and plot the line on the figure of part (a).

1-17. Would you expect D_{ij} to be a constant, as in a cubic lattice, for a liquid; for a glass? Justify your answers.

1-18. The general equations for diffusion in a two-dimensional lattice are:

$$J_x = -D_{11}\frac{dc}{dx} - D_{12}\frac{dc}{dy}$$

$$J_y = -D_{21}\frac{dc}{dx} - D_{22}\frac{dc}{dy}$$

Show what elements of the diffusion tensor D_{ij} are zero and which are equal for:

(a) A square array of points.

(b) A rectangular array of points.

This treatment should be valid for the diffusion of tracers *on* the surface of a crystal. If the exposed surface plane is an (hkl) plane, find the elements of D_{ij}.

chapter 2 ATOMIC THEORY

OF DIFFUSION

If a drop of a dilute mixture of milk in water is placed under a microscope and observed by transmitted light, small fat globules can be seen. These globules are about 1μ in diameter and continually make small movements hither and yon. These movements, which are called Brownian motion, give a continual mixing and are the cause, or mechanism, of the homogenization, whose rate could be measured in a macroscopic diffusion experiment. For example, if a drop of the same milky solution is placed in water, it will tend to spread out, and the mixture will ultimately become homogeneous. In this latter experiment a concentration gradient is present, a flux of fat globules[1] exists, and a diffusion coefficient could be measured. This is not quite an after-lunch experiment though, since turbulent mixing must be avoided and diffusion occurs quite slowly ($D \simeq 10^{-8}$ cm²/sec).

Brownian motion is not peculiar to the fat droplets in milk; in fact, active study at the turn of the century showed that it occurred for any microscopic particles suspended in any liquid or gas. This being the case, there is the interesting and potentially complicated question of how the random motion of these particles is related to the macroscopic displacement of the particles. For example, given the

---

[1] The composition of milk is not simple, but for our purpose it can be considered a colloidal dispersion of fat globules in water. Milk is used as an example because it is the most easily obtained dispersion with particles that can be resolved at 500×.

number of jumps per second and the mean jump distance, how far will the particle be from an arbitrary starting point after some very large number of jumps? This particular problem was initially treated about 1905 by Smoluchowski and by Einstein and has been further developed over the years.[1] It is generally called the random-walk problem.

This may seem like a digression from diffusion in crystalline solids, but the problems have striking similarities. It is impossible to observe the motion of the individual atoms in solids, but diffusion does occur in them, so there must be relative motion of the atoms. It is therefore reasonable to assume that diffusion occurs by the periodic jumping of atoms from one lattice site to another. If this is indeed true, then the mathematics of the random-walk problem will allow us to go back and forth between the observed macroscopic diffusion coefficients and the jump frequencies and jump distances of the diffusing atoms. The problem is not a simple one, but it is most exciting since it transforms the study of diffusion from the question of how fast a system will homogenize into a tool for studying the atomic processes involved in a variety of reactions in solids, for studying defects in solids, or for studying the interaction between the atoms themselves.

In this chapter we shall discuss the random-walk problem, the atomic mechanisms which are thought to give rise to diffusion, the factors which influence the jump frequency of the atoms, and the calculation of a diffusion coefficient from the combination of all of these.

2-1. RANDOM MOVEMENT AND THE DIFFUSION COEFFICIENT

Before discussing the detailed mechanisms and mathematics of diffusion, it is helpful to study a simple situation in which no detailed mechanism is assumed. In this section we shall derive an approximate equation relating D to the jump frequency and the jump distance without going through a rigorous treatment of the random-walk problem.

Consider a crystalline bar that has a concentration gradient along its y axis (see Fig. 2-1). We assume that the diffusing atoms can

---

[1] A readable, interesting treatment of Brownian motion can be found in the translation of Einstein's original papers. (A. Einstein, "Investigations on the Theory of Brownian Movement," Dover Publications, New York, 1956.) For an advanced, complete treatment which deals more with mathematics and less with physical phenomena, see N. Wax (ed.), "Selected Papers on Noise and Stochastic Processes," Dover Publications, New York, 1954.

only jump to the left or right and that when they do change position they jump a distance α along the y axis.   Consider now two adjacent lattice planes, designated 1 and 2, a distance α apart.   Let there be n_1 diffusing atoms per unit area in plane 1 and n_2 in plane 2.   If each atom jumps an average of Γ times per second, the number of atoms jumping out of plane 1 in the short period δt is $n_1\Gamma\delta t$.   Since half of these will go to the right to plane 2, the number of atoms jumping from plane 1 to plane 2 in δt is

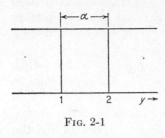

Fig. 2-1

$\frac{1}{2}n_1\Gamma\delta t$.   Similarly the number of atoms jumping from plane 2 to plane 1 in δt is $\frac{1}{2}n_2\Gamma\delta t$.   The net flux from planes 1 to 2 is thus

$$J = \frac{1}{2}(n_1 - n_2)\Gamma = \frac{\text{number of atoms}}{(\text{area})\ (\text{time})}$$

The quantity $(n_1 - n_2)$ can be related to the concentration or number per unit volume by observing that $n_1/\alpha = c_1$ and $n_2/\alpha = c_2$.   Thus

$$J = \tfrac{1}{2}(c_1 - c_2)\alpha\Gamma$$

But in essentially all diffusion studies c changes slowly enough with composition that

$$c_1 - c_2 = -\alpha\frac{\partial c}{\partial y}$$

so that
$$J = -\frac{1}{2}\alpha^2\Gamma\frac{\partial c}{\partial y} \qquad (2\text{-}1)$$

This equation is identical to Fick's first law if the diffusion coefficient D is given by

$$D = \tfrac{1}{2}\alpha^2\Gamma \qquad (2\text{-}2)$$

The diffusion coefficient is therefore determined by the product of the jump distance squared and the jump frequency.

It should be emphasized that Γ was assumed to be the same on planes 1 and 2 and the same for jumps to the left and to the right. Thus the flow down the concentration gradient does not result from any bias of the atoms to jump in that direction.   If each atom jumps randomly and $n_1 > n_2$, there will be a net flux from 1 to 2 simply because there are more atoms on plane 1 to jump to 2 than there are atoms on 2 to jump to 1.

Without assuming a particular mechanism, it is plausible that α is about the interatomic distance in a lattice, or the order of one angstrom. If we assume this, the jump frequency can be estimated from the measured diffusion coefficient. For carbon in α-Fe at 900°C, $D \simeq 10^{-6}$ cm²/sec. If $\alpha \simeq 10^{-8}$ cm, then $\Gamma \simeq 10^{10}$ sec⁻¹. That is, each carbon atom changes position about 10 billion times per second.

Near their melting points most fcc and hcp metals have a self-diffusion coefficient of 10^{-8} cm²/sec. Again taking $\alpha \simeq 10^{-8}$ cm gives $\Gamma \simeq 10^8$ sec⁻¹. Thus in most solid metals near their melting point each atom changes its site 100 million times a second. If this number seems impossibly large, remember that the vibrational frequency (Debye frequency) of such atoms is 10^{12} to 10^{13} sec⁻¹, so that the atom only changes position on one oscillation in 10^4 or 10^5. Thus, even near the melting point, the great majority of the time the atom is oscillating about its equilibrium position in the crystal.

2-2. MECHANISMS OF DIFFUSION

It is well known from the theory of specific heats that atoms in a crystal oscillate around their equilibrium positions. Occasionally these oscillations become violent enough to allow an atom to change sites. It is these jumps from one site to another which give rise to diffusion in solids. The discussion given in the preceding section and most of what is given below in this chapter will apply to several or all of the possible diffusion mechanisms. However, to aid the reader in understanding the applicability of the subsequent material, this section will be devoted to cataloguing the mechanisms which are thought to give rise to diffusion in crystalline solids.

Interstitial Mechanism. An atom is said to diffuse by an interstitial mechanism when it passes from one interstitial site to one of its nearest-neighbor interstitial sites without permanently displacing any of the matrix atoms. Figure 2-2 shows the interstitial sites of an fcc lattice. It should be noted that these sites also form an fcc lattice. An atom would diffuse by an interstitial mechanism in this lattice by jumping from one site to another on this sublattice of interstitial points.

To look at this process more closely, consider the atomic movements which must occur before a jump can occur. Figure 2-3 shows an interstitial atom in the (100) plane of a group of spheres packed into an fcc lattice. Before the atom labeled 1 can jump to the nearest-neighbor site 2 the matrix atoms labeled 3 and 4 must move apart enough to let it through. Actually if 1 rises out of the plane of the paper slightly as it starts toward 2, there is a partially formed channel available. Nevertheless an appreciable local dilatation of the lattice must occur before the jump can occur. It is this dilatation or distortion which constitutes the barrier to an interstitial atom changing sites. The basic problem in calculating a jump frequency is determining how often this barrier can be surmounted.

The interstitial mechanism is thought to operate in alloys for those solute atoms which normally occupy interstitial positions, e.g., C in α- and γ-iron. It will be dominant in any nonmetallic solid in which the diffusing interstitial does not distort the lattice too much. If the diffusing interstitial atom is almost as large as the atoms on the normal lattice site, then the distortion involved in this mechanism becomes too large, and another diffusion mechanism becomes dominant.

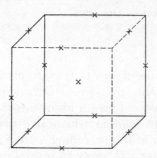

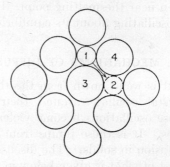

Fig. 2-2. x indicates the interstitial sites in an fcc unit cell.

Fig. 2-3. (100) plane in fcc lattice showing path of interstitial atom diffusing by interstitial mechanism.

Vacancy Mechanism. In all crystals some of the lattice sites are unoccupied. These unoccupied sites are called vacancies. If one of the atoms on an adjacent site jumps into the vacancy, the atom is said to have diffused by a vacancy mechanism.

Figure 2-4 shows the nature of the constriction which inhibits motion of an adjacent atom into a vacancy in an fcc lattice. If the undistorted lattice is taken to consist of close-packed spheres of diameter d, the distance between the restraining atoms (atoms labeled 1 and 2 in Fig. 2-4a) is $0.73d$. The distortion required to move an atom is thus small. In fact, the distortional energy put into the lattice in moving an iron atom into an adjacent vacancy is roughly equal to the energy required to move a carbon atom from one interstitial site to another in the same fcc phase. The reason that iron diffuses so much more slowly than carbon is that while each carbon atom always has several vacant nearest-neighbor sites, the fraction of vacant iron sites is very small, and each iron atom must wait an appreciable period before a vacancy becomes available.

The vacancy mechanism is well established as the dominant mechanism of diffusion in fcc metals and alloys and has been shown to be

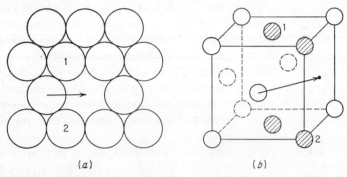

(a) (b)

FIG. 2-4. Two models showing the movement of an atom into an adjacent vacancy in an fcc lattice. (a) A close-packed plane of spheres. (b) A unit cell showing the four atoms (shaded) which must move before the indicated jump can occur.

operative in many bcc and hcp metals, as well as in ionic compounds and oxides.

Interstitialcy and Crowdion Mechanisms. Solute atoms which go into solution in metals as interstitials are appreciably smaller than the matrix atoms and, as discussed above, diffuse by the interstitial mechanism. If a relatively large atom such as a solvent atom gets into an interstitial position, how will it move? If it moves by an interstitial mechanism, it will produce a very large distortion in making a jump. Jumps which produce very large distortions occur infrequently, and so another diffusion mechanism which will require less distortion must be sought.

One of the mechanisms proposed for this purpose is the interstitialcy mechanism. Consider the interstitial atom shown in Fig. 2-5. It is said to diffuse by an interstitialcy mechanism if it pushes one of its nearest-neighbor atoms into an interstitial position and occupies the

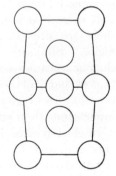

FIG. 2-5. (100) plane of fcc atom with interstitial atom on interstitial site.

lattice site previously occupied by the displaced atom. The distortion involved in this displacement is quite small, so it can occur with relative ease. The interstitialcy mechanism has been proved to be the dominant diffusion mechanism for silver in AgBr.[1] In this case the

[1] R. J. Friauf, *Phys. Rev.*, **105**: 843 (1957).

silver ion is smaller than the Br, and an interstitial silver ion does not distort the lattice unduly.

In the case of pure fcc metals the atoms are all the same size, and the distortion associated with the configuration shown in Fig. 2-5 is

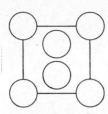

quite large. It has been shown that for Cu, and probably for all fcc metals, the arrangement of Fig. 2-6 has a lower energy.[1] Here two atoms share one site. Diffusion can occur by the rotation of this pair into another cubic direction or by the displacement of one of the atoms so that there are two atoms on one of the nearest-neighbor sites. This mechanism currently has no distinguishing name.

Fig. 2-6. (100) plane of fcc lattice with two atoms sharing one site. This is possible interstitial configuration in addition to that of Fig. 2-5.

Still another interstitial configuration is called the crowdion. It has the extra atom placed in a close-packed direction, thus displacing several atoms from their equilibrium position (see Fig. 2-7). This configuration resembles an edge dislocation in that its distortion is spread out along a line, it can glide in only one direction, and the energy to move it is quite small.

This multiplicity of configurations for an interstitial atom has prompted Lomer[2] to point out that an interstitial atom means only that there is one more atom than there are sites in a given small region. Similarly, he emphasizes that a vacancy need not mean that a particular site is vacant but that the region contains one fewer atom than sites.

In close-packed metals the energy required to form an inter-

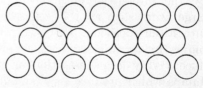

Fig. 2-7. (111) plane of fcc lattice showing a crowdion. (Note extra atom in middle row.)

stitial atom is so large that the concentration is negligible in any annealed sample. However, they may be formed in specimens by bombarding with high-energy particles or by cold working, and it is in annealing studies of such specimens that the above mechanisms are discussed.

[1] J. Gibson, A. Goland, M. Milgram, and G. Vineyard, *Phys. Rev.*, **120**: 1229 (1960).

[2] W. M. Lomer, in B. Chalmers (ed.), "Progress in Metals Physics," vol. 8, p. 255, Pergamon Press, Inc., New York, 1959.

Ring Mechanism. In the 1930s it was thought that self-diffusion in metals and alloys occurred by a simple exchange of two nearest neighbors. However, this mechanism requires distortions comparable to an interstitial mechanism for a solvent atom. By the late 1940s, the implications of this distortion and the Kirkendall effect (to be discussed in Chap. 4) had convinced most workers that this was not a likely mechanism. However, in 1950 Zener[1] pointed out that the distortion of the simple exchange could be appreciably reduced if, instead of two atoms interchanging, three or four atoms rotate as a group. The plausibility of this can be seen by looking at Fig. 2-8 and comparing the relative distortion that would result if an atom moved

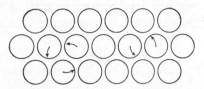

FIG. 2-8. Consider the much larger distortion involved in rotating two atoms as a group as opposed to rotating three atoms.

by a two-atom exchange and by a rotation of a three-atom ring. The strain induced by the rotation of a ring of four can best be seen with the help of a diagram similar to Fig. 2-3.

The ring mechanism is not known to operate in any metal or alloy, but it has been suggested as a mechanism which would explain some apparent anomalies in D for bcc metals. This is plausible due to the open nature of the bcc lattice.

2-3. RANDOM-WALK PROBLEM

After cataloguing the possible diffusion mechanisms, we turn next to the problem of relating these atomic jumps to the observed macroscopic diffusion phenomena. It has already been shown that near the melting point of many metals each atom changes sites roughly 10^8 times per second. Over the period of hours or days, the number of jumps becomes astronomical. These jumps are made in all directions and follow no particular pattern. Our problem is to take this welter of jumps and to calculate how far an atom will be from its initial site after n jumps. A first impression might be that the problem is insoluble, owing to the randomness of the atom's jumps, and indeed the exact distance cannot be calculated for a particular atom. However, precisely because of the random nature of the process and the large number of jumps, it is relatively easy to calculate the average distance that a group of atoms will have migrated from their initial

[1] C. Zener, *Acta Cryst.*, **3**: 346 (1950).

sites. This kind of problem is called a "random-walk" problem, and diffusion in crystalline solids is only one application of a broad group which includes the flipping of coins, the structure of polymers, and the theory of galaxies.

We will start with a general equation and make restrictions only as needed. Imagine an atom starting from the origin and making n jumps. The vector connecting the origin and the final position of the atom will be designated $\mathbf{R}_n$ and is given by the equation

$$\mathbf{R}_n = \mathbf{r}_1 + \mathbf{r}_2 + \mathbf{r}_3 + \cdots = \sum_{i=1}^{n} \mathbf{r}_i \qquad (2\text{-}3)$$

where the $\mathbf{r}_i$ are vectors representing the various jumps. To obtain the magnitude of $\mathbf{R}_n$, we square both sides of Eq. (2-3).

$$\begin{aligned}
\mathbf{R}_n \cdot \mathbf{R}_n = R_n{}^2 = {} & \mathbf{r}_1 \cdot \mathbf{r}_1 + \mathbf{r}_1 \cdot \mathbf{r}_2 + \mathbf{r}_1 \cdot \mathbf{r}_3 + \cdots + \mathbf{r}_1 \cdot \mathbf{r}_n \\
{} & + \mathbf{r}_2 \cdot \mathbf{r}_1 + \mathbf{r}_2 \cdot \mathbf{r}_2 + \mathbf{r}_2 \cdot \mathbf{r}_3 + \cdots + \mathbf{r}_2 \cdot \mathbf{r}_n \\
& \cdots\cdots\cdots\cdots\cdots\cdots\cdots\cdots\cdots\cdots\cdots\cdots \\
{} & + \mathbf{r}_n \cdot \mathbf{r}_1 + \mathbf{r}_n \cdot \mathbf{r}_2 + \mathbf{r}_n \cdot \mathbf{r}_3 + \cdots + \mathbf{r}_n \cdot \mathbf{r}_n \quad (2\text{-}4)
\end{aligned}$$

We can rewrite this array as a series of sums in which the first sum is the sum of the diagonal terms, $\Sigma \mathbf{r}_i \cdot \mathbf{r}_i$. The second sum will consist of all the terms $\mathbf{r}_i \cdot \mathbf{r}_{i+1}$ and $\mathbf{r}_{i+1} \cdot \mathbf{r}_i$. There are $n-1$ of each of these terms, and they can be said to lie along the semidiagonals of Eq. (2-4). Since $\mathbf{r}_i \cdot \mathbf{r}_{i+1}$ equals $\mathbf{r}_{i+1} \cdot \mathbf{r}_i$, these two sums can be combined. Proceeding in this manner gives

$$\begin{aligned}
R_n{}^2 &= \sum_{i=1}^{n} \mathbf{r}_i \cdot \mathbf{r}_i + 2 \sum_{i=1}^{n-1} \mathbf{r}_i \cdot \mathbf{r}_{i+1} + 2 \sum_{i=1}^{n-2} \mathbf{r}_i \cdot \mathbf{r}_{i+2} + \cdots \\
&= \sum_{i=1}^{n} r_i{}^2 + 2 \sum_{j=1}^{n-1} \sum_{i=1}^{n-j} \mathbf{r}_i \cdot \mathbf{r}_{i+j} \qquad (2\text{-}5)
\end{aligned}$$

To put this in the form we shall finally work with, note that by definition $\mathbf{r}_i \cdot \mathbf{r}_{i+j} = |\mathbf{r}_i|\,|\mathbf{r}_{i+j}| \cos \theta_{i,i+j}$ where $\theta_{i,i+j}$ is the angle between the two vectors. Substituting this relation in Eq. (2-5) gives

$$R_n{}^2 = \sum_{i=1}^{n} r_i{}^2 + 2 \sum_{j=1}^{n-1} \sum_{i=1}^{n-j} |\mathbf{r}_i|\,|\mathbf{r}_{i+j}| \cos \theta_{i,i+j} \qquad (2\text{-}6)$$

Please note that in the derivation of this equation no assumptions have been made concerning (1) the randomness of the jumps, (2) the lengths of the successive jumps, (3) the allowable values of $\theta_{i,i+j}$, or (4) the number of dimensions in which the atom is jumping. We shall now proceed to make assumptions about these and calculate an average value of R_n^2.

The problem of primary interest is that of diffusion in a crystalline solid. For crystals with cubic symmetry all the jump vectors will be equal in magnitude, and Eq. (2-6) can be written

$$R_n^2 = nr^2 + 2r^2 \sum_{j=1}^{n-1} \sum_{i=1}^{n-j} \cos \theta_{i,i+j}$$

$$= nr^2 \left(1 + \frac{2}{n} \sum_{j=1}^{n-1} \sum_{i=1}^{n-j} \cos \theta_{i,i+j} \right) \qquad (2\text{-}7)$$

This equation gives R_n^2 for one particle after n jumps. To obtain the average value of R_n^2, we must consider many atoms, each of which has taken n jumps. The quantity nr^2 will be the same for each flight, but the values of R_n^2 will be different, and the differences will arise from the differences in the magnitudes of the double sums. The average value of R_n^2 can be obtained by adding the various R_n^2 and dividing the sum by the number of atoms involved. The result can be written

$$\overline{R_n^2} = nr^2 \left(1 + \frac{2}{n} \overline{\sum_{j=1} \sum_{i=1} \cos \theta_{i,j+j}} \right) \qquad (2\text{-}8)$$

If each jump direction is independent of the direction of the jumps which preceded it and each jump vector and its negative are equally probable, then positive and negative values of any given $\cos \theta_{i,i+j}$ will occur with equal frequency, and the average value of the term involving the double sum will be zero. When this is true

$$\overline{R_n^2} = nr^2 \qquad (2\text{-}9)$$

or $$\sqrt{\overline{R_n^2}} = \sqrt{n}\, r \qquad (2\text{-}10)$$

Two aspects of this result are particularly noteworthy. The first is the extreme simplicity of the equation. The second is the fact that the mean displacement (actually the root-mean-square displacement) is proportional to the square root of the number of jumps.

As a simple example of the step between Eqs. (2-8) and (2-9), consider the case of a single atom jumping back and forth along a line. The values of $\cos \theta_{i,i+j}$ are now either $+1$ or -1 since the angle between any two jump vectors will be either $0°$ or $180°$. If we consider many lines, each with a particle initially at $x = 0$ as in Fig. 2-9a, after n jumps the particles will be distributed at various distances from $x = 0$ as shown in Fig. 2-9b. From this figure it is apparent that $R_n{}^2$ for the various particles differs appreciably. However, since values of $\cos \theta = +1$ and $\cos \theta = -1$ are equally probable, $2/n$ times the average value of $\cos \theta$ for all jumps will be much less than 1.

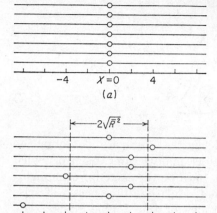

(a)

(b)

FIG. 2-9. (a) Initial distribution of atoms, one to a line. (b) Final distribution after each atom took 16 random jumps. $\sqrt{\overline{R^2}}$ is the calculated root mean square for the points shown.

Although the number of jumps and number of atoms are very small, let us consider in more detail the case depicted in Fig. 2-9. To obtain the path of each atom, a coin was flipped 16 times. A head represented a jump to the right; a tail, a jump to the left. In this way successive jumps were independent of each other, and jumps to the left and to the right were equally probable. The root-mean-square displacement of this group of atoms is found to be 3.2 units. Since there were 16 jumps per atom, Eq. (2-9) predicts a value of 4.0, which is in reasonable agreement considering the very small number of jumps involved.

To give a better understanding of the effect of the random movement of atoms for very large values of n, consider the case of carbon diffusing in γ-iron. At an average carburizing temperature (950°C) carbon atoms make 10^{10} jumps per second. Taking the jump distance to be 1 A, in 1 sec each carbon atom travels a *total* distance of 1 m and a *net* distance of 10^{-3} cm. After the 10^4 sec (about 3 hr) of an average carburizing treatment, the mean penetration is 0.10 cm, while the total distance traveled by the atom is 10 km, or roughly 6 miles. It is thus obvious that the net displacement of each atom is extremely small compared to the total distance it travels.

Random Walk in Three Dimensions. As a second example of what is involved in going from Eq. (2-8) to Eq. (2-9), consider the case of a vacancy diffusing in an fcc lattice.[1]   Figure 2-10 shows the orientation

[1] The proof given here applies equally well for any other cubic lattice, in which all atoms are on equivalent sites.

of the 12 possible jump vectors for this lattice. All of these vectors are equivalent so that each will occur with the same frequency if an average is taken over many vacancies and a very large number of jumps. For this reason the mean value of the quantity $\mathbf{r}_i \cdot \mathbf{r}_{i+j}$, that is, $\overline{\mathbf{r}_i \cdot \mathbf{r}_{i+j}}$ $(i \neq j)$, is independent of the vector chosen for $\mathbf{r}_i$. Furthermore, the very large number of vectors to be dotted into $\mathbf{r}_i$ will consist of the 12 vectors of Fig. 2-10 in equal proportions. Thus the mean value of the double sum in Eq. (2-5) will be zero if

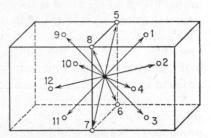

FIG. 2-10. The arrows show 12 possible jump vectors in an fcc lattice.

$$\sum_{j=1}^{12} \mathbf{r}_i \cdot \mathbf{r}_j = r^2 \sum_{j=1}^{12} \cos \theta_{ij} = 0 \qquad (2\text{-}11)$$

where the summation of j ranges over the 12 jump vectors of Fig. 2-10. It can be seen in Fig. 2-10 that for any particular jump vector there is another jump vector equal to the negative of that vector. For example, $\mathbf{r}_5 = -\mathbf{r}_7$. Therefore

$$\mathbf{r}_i \cdot \mathbf{r}_7 + \mathbf{r}_i \cdot \mathbf{r}_5 = \mathbf{r}_i \cdot \mathbf{r}_7 - \mathbf{r}_i \cdot \mathbf{r}_7 = 0$$

Pairing up each of the 12 vectors with its negative in this way, we see that Eq. (2-11) is satisfied, and so again we obtain

$$\overline{R_n^2} = nr^2 \qquad (2\text{-}9)$$

A similar argument can be given for the case of liquids or gases. The two main differences between this case and crystals is that all values of $\theta_{i,i+j}$ are possible instead of a discrete set and all values of $\mathbf{r}_i$ are possible (although in any given system the values will cluster about some mean). If the jumps, or flights, $\mathbf{r}_j$ and $-\mathbf{r}_j$ are equally probable, they occur with the same frequency, and the double sum again goes to zero. The resulting equation is

$$\overline{R_n^2} = \sum_{i=1}^{n} r_i^2 = n\overline{r^2} \qquad (2\text{-}12)$$

This equation differs from Eq. (2-9) only in that the unique jump distance of the crystal is replaced by a root-mean-square jump distance.

Relation of D to Random Walk. There are several ways of deriving the equation relating D to the atomic jump frequency and jump distance.   One is to consider an atom starting from the origin, as in the preceding section, but instead of calculating $\overline{R_n^2}$ to calculate the probability that the atom is between r and $r + dr$ after n jumps.   This probability can also be calculated by solving the same problem using Fick's laws.   In addition to giving the desired equation for r, this procedure shows that the solutions are completely equivalent.   The main drawback to this approach for our purposes is that the probability problem is complicated, though not difficult.[1]

A second, simpler approach is to consider the same problem but to solve only for $\overline{R_n^2}$ using the two techniques.   It is shown in one of the problems at the end of this chapter that, if $\Gamma \equiv n/t$ and α is the jump distance,

$$\overline{R_n^2} = n\alpha^2 = 6Dt \tag{2-13}$$

so that
$$D = \tfrac{1}{6}\Gamma\alpha^2 \tag{2-14}$$

The difference of a factor of 3 between this and Eq. (2-2) comes from the restriction of the previous derivation to jumps along one direction. Thus, only one-third of the jumps taking place in three dimensions are along the axis considered in Sec. 2-1.

A still simpler method, which is similar to that of Sec. 2-1, will be given here.   The main virtue of this procedure is that the simple mathematics cannot come between the reader and the physical model being studied.   In this way the assumptions can be emphasized as they are made.   The difference between this derivation and that of Sec. 2-1 is that we now wish to work with a particular mechanism in a particular three-dimensional lattice.   Since most of the self-diffusion studies have been made on fcc metals, we shall consider the diffusion of a tracer in a pure fcc metal by a vacancy mechanism.   We wish to calculate the tracer flux along the x axis.

If Γ is the average number of jumps per second for each tracer atom and n_1 is the number of tracer atoms on plane 1, $n_1\Gamma\delta t$ of the tracer atoms on plane 1 will jump in the short period δt.   The quantity $\Gamma\delta t$ will be proportional to the number of nearest-neighbor sites, to the probability that any given neighboring site is vacant (p_v), and to the probability that the tracer will jump into a particular vacant site in δt, namely $w\delta t$.[2]   Thus we can write

$$\Gamma\delta t = 12p_v w\delta t \tag{2-15}$$

---

[1] A solution of this type is given in the first few pages of B. S. Chandrasekhar, *Revs. Modern Phys.*, **15**: 1 (1943).   This article is also reprinted in N. Wax (ed.), "Selected Papers on Noise and Stochastic Processes," Dover Publications, New York, 1954.

[2] If Γ_v is the number of jumps a vacancy makes per second, then $12w = \Gamma_v = \Gamma/p_v$.

Since only 4 of the 12 nearest neighbors are on plane 2 (see Fig. 2-10), the flux per unit area from plane 1 to plane 2 is

$$J_{12} = 4n_1 p_{v2} w_{12}$$

where subscripts have been added to emphasize the planes involved; for example, p_{v2} is the probability that any site on plane 2 is vacant. Similarly the reverse flux is given by

$$J_{21} = 4n_2 p_{v1} w_{21}$$

In alloys, w and p_v will change with composition and thus give rise to a variation of D with composition. However, all isotopes of a metal are assumed to act the same, so that in a pure metal $w_{12} = w_{21}$ and $p_{v1} = p_{v2}$. Substituting these, and $n_1 = \alpha c_1 = (a_0/2) c_1$, gives for the net flux

$$J = 4\alpha p_v w(c_1 - c_2) \tag{2-16}$$

or substituting

$$c_1 - c_2 = -\frac{a_0}{2} \frac{\partial c}{\partial x}$$

we get

$$J = -a_0^2 p_v w \frac{\partial c}{\partial x}$$

From the assumed equivalence of all isotopes of the same metal, it follows that p_v will be equal to the fraction of sites vacant, or N_v. The desired equation is then

$$D = a_0^2 N_v w \tag{2-17}$$

The calculation of D in a pure fcc metal is thus reduced to the problem of calculating the mole fraction of vacancies and the jump frequency of an atom into an adjacent vacancy. Conversely, if we measure D, since a_0 is known, we can calculate $N_v w$; or knowing N_v, we can calculate w.

In a derivation similar to that which led to Eq. (2-17), it is possible to derive an equation for an interstitial solute in a binary alloy. If the solution is very dilute, w is independent of composition, and the mole fraction of vacant sites is essentially unity. Thus for a very dilute alloy, D for the interstitial element is

$$D = \gamma a_0^2 w \tag{2-18}$$

where γ is a geometric constant derivable from Eq. (2-14).

In closing, note that both Eqs. (2-17) and (2-18) can also be derived from Eq. (2-14) by substituting $\alpha = a_0/\sqrt{2}$ and $\Gamma = 12wN_v$. The advantage of the derivation given above is its relative simplicity and the ease with which it can be extended to the cases in which p_v and w vary with composition. However, once the assumptions leading to Eqs. (2-17) and (2-18) are clearly in mind, it is usually easier to work with Eq. (2-14).

2-4. CALCULATION OF D

Our study of the atomistic processes contributing to diffusion has led to Eq. (2-17). From this point on, our understanding of D will increase in proportion to our understanding of w and N_v. Thus the calculation of these terms is one of the basic problems in the atomic approach to diffusion. In this section we shall review the methods of evaluation, trying to emphasize the assumptions and approximations involved. The discussion to be given is strictly applicable only to self-diffusion in pure metals or interstitial diffusion in very dilute, binary alloys. The changes needed in an extension to substitutional alloys and more concentrated interstitial alloys will be discussed in Chap. 4.

Equilibrium Concentration of Vacancies. To gain a better understanding of D for a vacancy mechanism, we consider first the problem of how many vacancies will be present in a pure metal and how this concentration will change with temperature. The most important concept to be grasped here is the increase in entropy which results from the mixing of two pure components. A plausibility argument for this increase in entropy can be seen from the following. If a drop of ink is placed in a glass of water, the mixture will ultimately become uniformly tinted. An explanation for this homogenization might be found in Fick's laws, but it could also be found in a basic thermodynamic requirement for equilibrium; namely, that for equilibrium the entropy of any isolated system will be a maximum. Thus, this homogenization or mixing of the ink-water mixture must correspond to an increase of the entropy of the mixture.

To be more quantitative, if an ideal solution is formed upon the addition of component 1 to component 2, the equation for the increase of entropy per mole of solution is

$$\Delta S_{\text{mix}} = -R[(1 - N_1) \ln (1 - N_1) + N_1 \ln N_1] \qquad (2\text{-}19)$$

where N_1 is the mole fraction of component 1.[1] This equation is

---

[1] The student who is not familiar with Eq. (2-19) is referred to A. Cottrell, "Theoretical Structural Metallurgy," p. 102, St. Martin's Press, Inc., New York, 1957, where it is derived for crystals.

plotted in Fig. 2-11.   It can be seen that the entropy per mole of any mixture is greater than that of the pure components.   To apply Eq. (2-19) to the study of vacancies in metals, consider a large piece of pure metal with no sites vacant.   If several vacancies are taken from the

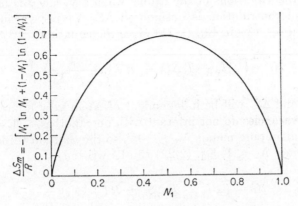

FIG. 2-11. ΔS_m is the entropy increase upon forming one mole of an ideal solution from the pure components.   The slope of the curve is infinite at the limits of $N_1 = 0$ and $N_1 = 1$.

surface and mixed throughout the metal, the increase in the entropy, per mole of solution $\delta \Delta S_{\text{mix}}$, is

$$\delta \Delta S_{\text{mix}} = \frac{d \Delta S_{\text{mix}}}{d N_v} \delta N_v = -R \ln \frac{N_v}{1 - N_v} \delta N_v \qquad (2\text{-}20)$$

where δN_v is the change in the mole fraction of vacancies.   In the limit of $N_v \to 0$ it can be seen that $\delta \Delta S_{\text{mix}}/\delta N_v \to \infty$.   That is, the increase in entropy per vacancy added is extremely large for the first few vacancies, but it continually decreases from its initial, infinite value.   It follows that there will always be some vacancies in any annealed piece of metal.   To calculate just what the equilibrium value of N_v will be, we use the fact that in any isothermal, isobaric system at equilibrium the change in the Gibbs free energy G will be zero for any small displacement.   If δn_v additional vacancies are mixed into a mole of a crystal already containing the concentration N_v of vacancies, the change in G will be

$$\delta G = \Delta H_v \frac{\delta n_v}{N} - T \frac{\partial S}{\partial N_v} \frac{\delta n_v}{N} \qquad (2\text{-}21)$$

where N is Avogadro's number. $\Delta H_v/N$ is the increase in the enthalpy of the crystal per vacancy added and stems from the local changes in the atomic and electronic configurations of the crystal when a vacancy is introduced. The increase in the entropy of the lattice per vacancy added, $(\partial S/\partial N_v)(1/N)$, arises from the ideal entropy of mixing as given by Eq. (2-20) and a second part which stems primarily from the change in the vibrations of the atoms when a vacancy is introduced. This second contribution is designated $\Delta S_v/N$† per vacancy. Substituting these terms in Eq. (2-21) gives the equation

$$\delta G = \left(\Delta H_v - T\,\Delta S_v + RT \ln \frac{N_v}{1 - N_v} \right) \frac{\delta n_v}{N} \tag{2-22}$$

Both ΔH_v and ΔS_v will be independent of N_v in very dilute solutions where the vacancies do not interact with one another. (Experiments indicate that in pure metals $N_v \lesssim 10^{-4}$, so the solution is indeed very dilute.) Since $N_v \ll 1$, Eq. (2-22) can be written

$$\delta G = (\Delta H_v - T\,\Delta S_v + RT \ln N_v) \frac{\delta n_v}{N} \tag{2-23}$$

But, at equilibrium $\delta G = 0$ for any small δn_v. Thus at equilibrium, N_v must have the value given by the equation

$$N_v{}^e = \exp \frac{\Delta S_v}{R} \exp \left(-\frac{\Delta H_v}{RT} \right) \tag{2-24}$$

where the superscript e is added to N_v to emphasize that $N_v{}^e$ is a particular value of N_v instead of a variable. This equation can also be written

$$N_v{}^e \equiv \exp \left(\frac{-\Delta G_v}{RT} \right) \tag{2-25}$$

where $\Delta G_v \equiv \Delta H_v - T\,\Delta S_v$ is the free-energy change of an infinite crystal, per mole of vacancies added, over and above the entropy of mixing. An equation identical in form to Eqs. (2-24) or (2-25) could be obtained for the concentration of interstitial metal atoms, N_i. In it, $\Delta G_i = \Delta H_i - T\,\Delta S_i$ would replace ΔG_v.

A physical feeling for the meaning of, and basis for, Eq. (2-25) can be obtained by studying Fig. 2-12. Here the molar decrease in the free energy per mole of vacancies added is given by the line $-RT \ln N_v$. The horizontal line represents the free-energy increase per mole of

---

† Strictly speaking, ΔS_v is a parameter which when added to the ideal entropy of mixing gives the observed entropy effect. In solution chemistry ΔS_v would be called an excess entropy of mixing.

vacancies added (ΔG_v). The system will adjust N_v until Eq. (2-24) is satisfied, that is, until $\Delta G_v = -RT \ln N_v$. If the temperature is suddenly increased to T_2, $RT \ln N_v$ will increase while ΔG_v will be essentially unchanged. In order to reestablish equilibrium, N_v will increase until $-RT \ln N_v$ again equals ΔG_v.

For vacancies in gold, $\Delta H_v \simeq 23.0$ kcal/mol ($\simeq 1.0$ ev per vacancy). Thus in the temperature range of 900–1000°C, N_v in gold will roughly double with a 90°C increase in temperature. Often authors write $N_v = \exp(\Delta H_v/RT)$, omitting the term including ΔS_v. This is not correct, but ΔS_v is not known in most cases. The few data available indicate that $\exp(\Delta S_v/R) \lesssim 10^1$. The omission of this term often gives an adequate approximation and avoids the problems of discussing ΔS_v. Taking $\Delta S_v \simeq 0$ and $\Delta H_v \simeq 23.0$ kcal/mole, we get $N_v \simeq 10^{-4}$ at 980°C.

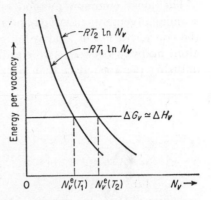

FIG. 2-12. $N_v^e(T)$, the equilibrium concentration of vacancies at the temperature T, is attained when $T(d\,\Delta S_{mix}/dN_v)$ equals $-\Delta G_v$. The variation of both of these quantities with N_v is shown here, at two different temperatures.

Calculation of the Jump Frequency w. The second unknown quantity which enters into D is w, the frequency with which an atom will jump into an adjacent, vacant site. The calculation of w, or even its temperature dependence, from our fundamental knowledge of the forces between atoms and reaction kinetics is a very difficult problem. Actually, our present knowledge is such that any calculation from fundamentals cannot give a real check on experimental results. The main purpose in such a theoretical study is to develop greater insight into the factors which determine w and thereby D.

The atom movements required for an atom to jump are shown schematically in Fig. 2-13; (a) and (c) show the initial and final states, while (b) shows the midway configuration referred to as the activated state. There are two separate requirements to be met before the group of atoms can go from (a) to (c). First, the diffusing atom must be moving to the right far enough to carry it into the vacant site; and second, the two restraining atoms must simultaneously move apart a great enough distance to let the diffusing atom through. Whenever

[1] R. Simmons and R. Balluffi, *Phys. Rev.*, **119**: 600 (1960).

these two steps occur at the same time, the diffusing atom will change sites.

The most common method of calculating w ignores the detailed atomic movements involved and uses statistical mechanics to calculate the concentration of "activated complexes," or regions containing an atom midway between two equilibrium sites. The number of atoms diffusing per second is then obtained by multiplying the number of

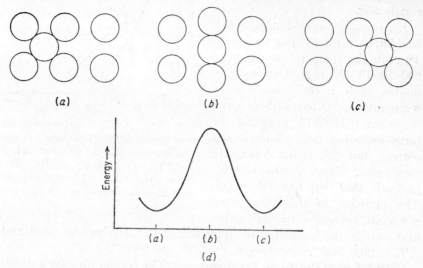

FIG. 2-13. (a), (b), and (c) are schematic drawings showing the sequence of configuration involved when an atom jumps from one normal site to a neighboring one. (d) shows how the free energy of the entire lattice would vary as the diffusing atom is reversibly moved from configuration (a) to (b) to (c).

activated complexes (n_m) by the average velocity of the atoms moving through this midpoint $\bar{v}$, divided by the width of the barrier or midpoint δ. From this number jumping per second, it is shown that the average jump frequency per atom is $w = N_m \bar{v}/\delta$, where N_m is the mole fraction of activated complexes. This result will be derived or rationalized here without the aid of statistical mechanics. This simplified treatment is less rigorous, but the basic assumptions are still apparent.[1]

[1] Those interested in a more rigorous derivation of these results should see C. Wert, *Phys. Rev.*, **79**: 601 (1950). The student not familiar with the rudiments of statistical mechanics will find E. Guggenheim, "The Boltzmann Distribution," North-Holland Publishing Company, Amsterdam, 1955, helpful. It is brief and lucid and emphasizes the relationship between thermodynamics and statistical mechanics. The most general and rigorous treatment using this approach.has been given by G. Vineyard, *J. Phys. Chem. Solids*, **3**: 121 (1957).

The diffusing atom shown schematically in Fig. 2-13b is said to be at the saddle point. Throughout the crystal there will always be atoms entering this configuration as well as leaving it. To calculate the number of atoms at the saddle point at any instant, it is necessary to know the increase in the Gibbs free energy of a region when an atom in it moves from a normal site to the saddle-point position, ΔG_m. Zener[1] suggested that this free-energy change could be visualized in the following thought experiment. If the diffusion direction is defined as the x axis, we constrain the atom so that it can execute its normal vibration only in the yz plane. The atom is then slowly moved from its initial site to the saddle point, allowing the surrounding atoms to continuously readjust their positions. The work done in this reversible, isothermal process, at constant pressure, is just equal to the change in Gibbs free energy for the region (ΔG_m). This can be written

$$\Delta G_m = \Delta H_m - T \, \Delta S_m \tag{2-26}$$

It is assumed that ΔG_m has all the properties possessed by ΔG_v of Eq. (2-25). Using this ΔG_m, the equilibrium mole fraction of atoms in the region of the saddle point N_m can be calculated using a treatment essentially the same as was used in obtaining the equation for $N_v{}^e$, that is, Eq. (2-25). Instead of mixing into the lattice vacancies which increase the free energy by ΔG_v per mole of vacancies, we mix in activated complexes which increase the free energy by ΔG_m per mole of complexes. The ideal entropy of mixing is the same for vacancies and complexes so, at equilibrium, n_m out of N atoms will be in the neighborhood of a saddle point at any instant and

$$\frac{n_m}{N} \equiv N_m = \exp \frac{- \, \Delta H_m + T \, \Delta S_m}{RT} = \exp \left(- \frac{\Delta G_m}{RT} \right) \tag{2-27}$$

In the equation $w = N_m \bar{v}/\delta$, simple dimensional analysis shows that $\bar{v}/\delta$ is a frequency. This is the frequency with which the atoms at the saddle point go to the new site v. A more complete treatment shows that v is of the order of the mean vibrational frequency of an atom about its equilibrium site. Thus, of N atoms $n_m v$ will jump from one site to a given neighbor site per second. If this is true, the average jump frequency for any given atom will be

$$\frac{n_m v}{N} = w = v \exp \left(- \frac{\Delta G_m}{RT} \right) \tag{2-28}$$

[1] C. Zener, in W. Schockley (ed.), "Imperfections in Nearly Perfect Crystals," p. 289, John Wiley & Sons, Inc., New York, 1952.

A particularly simple interpretation of Eq. (2-28) is to think of it as the frequency with which an atom vibrates in a given diffusion direction ν times the probability that any given oscillation will move the atom to an adjacent site in that direction, $\exp(-\Delta G_m/RT)$. The precise definition of ν is one of the more difficult aspects of a rigorous theory. However, it is usually taken equal to the Debye frequency.

Before proceeding, let us discuss the assumptions which enter into the derivation of Eq. (2-28). The entire treatment is premised on the argument that one can use equilibrium thermodynamics to relate ΔG_m and N_m. For example, Zener's definition of ΔG_m given above assumes that it is a state function, that is, that the free energy of the activated complex attained by the reversible process is the same as the free energy of the activated complex that really occurs in nature. This would be true if the complex, like a vacancy, existed for a period long enough to let the surrounding lattice completely adjust to the presence of the complex. If the complex does not exist for long enough to reach equilibrium with the surrounding lattice, one might be tempted to say that this could be corrected for by simply taking a slightly larger value of ΔG_m than that given by the thought experiment. However, this is a dangerous solution since in this limit the statistical mechanical procedure which forms the entire basis for the treatment may not be applicable. In a more rigorous treatment it is assumed that the probability of a given state occurring is the probability that would be valid for a system in thermal equilibrium; but if the system is not in thermal equilibrium, the entire procedure may become only an approximation of doubtful value.

One of the main justifications of this analysis is that *it works*. It is not clear to many authors just how the assumption discussed above can be fulfilled, but whenever the predictions of this theory have been experimentally checked, the theory has been vindicated.

It was pointed out initially that before an atom could change sites it had to move to the saddle-point position at the same time that the restraining atoms had moved out of the way. Rice has developed a formalism for treating this problem which is based on the actual vibrational modes of the atoms.[1] He assumes that the movement of a particular atom will be that which results from the superposition of the displacements due to the elastic waves in the region of the atom involved. If the various waves occur with random phases, they will occasionally superimpose to give the displacements required for a jump. Using this analysis, an equation identical in form to Eq. (2-28) is

[1] S. Rice, *Phys. Rev.*, **12**: 804 (1958).

obtained. However, in this case the quantities which replace ν and ΔG_m are expressed in terms which can be explicitly summed.

This procedure avoids the requirement that the activated complex be in thermal equilibrium with the lattice. However, this does not mean that we can now calculate w with any certainty. To simplify the complicated mathematics involved, it is assumed that the forces between atoms are harmonic. For the large displacements involved this is not a good approximation. And, even with this simplification, the actual solving of the equations for a particular material would be very complicated. Because of these difficulties, the contribution of Rice's procedure is more conceptual than practical. Manley[1] has generalized Rice's treatment and shown that, to a first approximation, the quantity which replaced ΔG_m in this approach is equal to the work required to move the atom into the saddle-point configuration, as was the case in the statistical mechanical treatment.

Equations for D. Empirically it is found that D can be described by the equation

$$D = D_0 \exp\left(-\frac{Q}{RT}\right) \tag{2-29}$$

where D_0 and Q may vary with composition but are independent of temperature. Experimentally D_0 and Q are obtained by plotting $\ln D$ versus $1/T$. The slope of this plot gives

$$\frac{d \ln D}{d\, 1/T} = -\frac{Q}{R}$$

while $\ln D_0$ is given by the intercept at $1/T = 0$.

An alternate equation for D in the case of interstitial diffusion can be obtained by substituting Eq. (2-28) for w in Eq. (2-18). This gives

$$D = \left(\gamma\, a_0{}^2 \nu \exp\frac{\Delta S_m}{R}\right) \exp\left(-\frac{\Delta H_m}{RT}\right) \tag{2-30}$$

Comparing this with Eq. (2-29) we see that the first term in parentheses is equal to D_0 and that Q equals the quantity ΔH_m.

For diffusion by a vacancy mechanism in a pure metal, Eqs. (2-28) and (2-17) give

$$D = \left[a_0{}^2 \nu \exp\left(\frac{\Delta S_f + \Delta S_m}{R}\right)\right] \exp\frac{-\Delta H_f - \Delta H_m}{RT} \tag{2-31}$$

The term in square brackets is again D_0, while Q is the sum of ΔH_m and ΔH_f. Since Q is seen to be made up of enthalpy terms in both cases, it will be replaced by ΔH in the rest of this book.

[1] O. Manley, *J. Phys. Chem. Solids*, **13:** 244 (1960).

The ΔS term from either Eq. (2-30) or Eq. (2-31) can be evaluated from the known value of D_0, $a_0{}^2$, γ, and an assumed value of ν. As was pointed out above, ν is usually taken to be the Debye frequency for a pure metal. In the case of interstitial atoms, ν can be calculated by assuming that the potential-energy curve of the atom varies sinusoidally along the diffusion path and its maximum value is ΔH_m.[1] In either case, the value of ΔS obtained depends on the value of ν assumed. In view of the vagueness as to what ν is to be used, ΔS cannot be determined with precision. However, as will be seen in the next section, the evaluation of even an approximate ΔS can be quite helpful in checking experimental results.

2-5. ZENER'S THEORY OF D_0

The quantity in D_0 which is least well known a priori is ΔS. In the next section we shall briefly discuss the calculation of ΔS_m and ΔS_v from "first principles," but here we shall give a semiempirical method due to Zener.[2] Experimentally, this procedure is useful for estimating D_0 where it is not known, while pedagogically it sheds light on the meaning of ΔG_m. Through its agreement with experiment it also lends credence to the assumptions leading to Eq. (2-28).

Interstitial Diffusion. Zener's treatment applies only to ΔS_m so the case of interstitial diffusion will be discussed first. In the reversible movement of the atom from its equilibrium position x_0 to the saddle point, x', Zener reasoned that much of the work goes into elastically straining the lattice around the saddle point. Thus the work ΔG_m can be approximated by the elastic strain set up in the lattice upon moving the diffusion atom from x_0 to x'.

$$\Delta G_m = \text{work} = \int_{x_0}^{x'} f \, dx \approx \mu \epsilon_0{}^2 \tag{2-32}$$

where ϵ_0 is some representative strain for the matrix when the atom is at the saddle point and where μ is an appropriate elastic modulus for the solvent. The force required (f) will not be strictly proportional to the strain, for strains as large as those encountered here, but Eq. (2-32) is basically reasonable, if not highly accurate. Since ΔG_m was assumed to be a Gibbs free energy, it follows that

$$\Delta S_m = -\left(\frac{d \, \Delta G_m}{dT}\right)_P \tag{2-33}$$

[1] C. Wert and C. Zener, *Phys. Rev.*, **76**: 1169 (1949).
[2] C. Zener, in W. Shockley (ed.), p. 289, "Imperfections in Nearly Perfect Crystals," J. Wiley, New York, 1952.

But ϵ_0 is essentially independent of temperature, so ΔG_m varies with temperature in the same way as μ. Experimentally it is found that $d\mu/dT$ is negative for all solids not undergoing a phase change; therefore, ΔS_m will be positive. Thus we can immediately set a lower limit on D_0 (interstitial) and write

$$D_0 \geq a_0{}^2\nu \simeq 10^{-3} \text{ cm}^2/\text{sec} \qquad (2\text{-}34)$$

This must be true if the above argument has any basis in fact.

To be more quantitative, define ΔG_0 as equal to ΔG_m at zero degrees absolute. Then

$$\Delta S_m = -\Delta G_0 \frac{d(\Delta G/\Delta G_0)}{dT} \qquad (2\text{-}35)$$

but $\Delta G_0 \equiv \Delta H_0$ and using Eq. (2-32) we get

$$\Delta S_m \simeq -\Delta H_0 \frac{d(\mu/\mu_0)}{dT} \qquad (2\text{-}36)$$

To simplify the application of this equation, Zener defined a new parameter β:

$$\beta \equiv -\frac{d(\mu/\mu_0)}{d(T/T_{\mathrm{mp}})}$$

where T_{mp} is the melting point of the solvent. The calculated value of β for a variety of metals ranges from 0.25 to 0.45, with 0.35 being a good average value. Also, it is an empirical fact that ΔH does not vary with temperature, so that $\Delta H \simeq \Delta H_0$ is a good approximation. Substituting β and ΔH in Eq. (2-36) gives Zener's final equation,

$$\Delta S_m \simeq \beta \frac{\Delta H_m}{T_{\mathrm{mp}}} \qquad (2\text{-}37)$$

Notice that the $\epsilon_0{}^2$ of Eq. (2-32) does not appear explicitly in this equation. Thus in a given solvent, β and T_{mp} are constant, and the equation predicts that ΔS_m is directly proportional to ΔH_m.

About 1950, when Zener put forth this analysis, the available data on D_0 for interstitials ranged over many orders of magnitude. The data were thus in marked disagreement with Eq. (2-37), which states that ΔS and thus D_0 should lie within a much narrower range. Table 2-1 summarizes some of the very accurate data which have been determined since 1950. All these newer data lie within the range of values predicted by Eq. (2-37). In Table 2-1, the agreement between ΔS_{exp} and ΔS_{th} [calculated from Eq. (2-37)] may not appear to be outstanding, and indeed it is not. However, the very accurate data which are now available have been almost entirely determined since 1950. The data available then were much poorer, and Zener's theory[1] was instrumental in casting doubt on the validity of much of the older work.

TABLE 2-1. Theoretical and Experimental ΔS for Interstitials

Solvent	Solute	Ref.	$D_0 \left(\dfrac{cm^2}{sec}\right)$	ΔH (kcal/mole)	$(\Delta S/R)_{th}$	$(\Delta S/R)_{exp}{}^a$
Ta	O	1	0.0044	25.45	1.6	0.79
	N	1	0.0056	37.8	2.3	0.73
	C	1	0.0061	38.5	2.4	0.73
Fe	C	2	0.020	20.1	2.2	2.4
	N	3	0.003	18.2	2.0	0.69
Nb	O	1	0.021	26.9	2.0^b	2.4
	N	1	0.0086	34.9	2.6^b	1.3
	C	1	0.0040	33.0	2.4^b	0.51

[1] R. Powers and M. Doyle, *J. Appl. Phys.*, **30**: 514 (1959).
[2] C. Wert, *Phys. Rev.*, **79**: 601 (1950).
[3] C. Wert, *J. Appl. Phys.*, **21**: 1196 (1950).
[a] Following Wert and Zener, ν was calculated from the equation $\nu = (\Delta H/2m\alpha^2)^{\frac{1}{2}}$ where m is the mass of the solute and α the jump distance.
[b] β assumed equal to 0.4.

As an example of the use of Eq. (2-37), consider the following rough calculation for carbon in α-Fe. Given that $\beta \simeq 0.4$, $\Delta H \simeq 20,000$ cal/mole, $T_{mp} \simeq 1800°K$, we obtain $\Delta S_{th} \simeq 2.2\,R$. Thus $\exp(\Delta S/R) \simeq 9$ and with $\nu \simeq 10^{13}$ sec^{-1}, $D_{0,th} \simeq 0.015$ cm^2/sec. Experimentally D_0 is found to be 0.020 cm^2/sec so that the agreement here is excellent.

Application of Zener's Theory to the Vacancy Mechanism. Zener's theory of D_0 works well for interstitial solutes, and thus it is desirable to extend it to the case of self-diffusion by the vacancy mechanism in pure metals. The argument leading to ΔS_m for vacancy diffusion is exactly the same as the above. The new aspect of the vacancy case comes in estimating ΔS_v. This cannot be done with precision, and Zener simply rationalized as follows. We can write $\Delta S_v \sim \ln(\nu_0/\nu_v)$† where ν_0 is the average vibrational frequency of an atom in a perfect crystal and ν_v is the average frequency of the atom after a vacancy

[1] *Ibid.*, p. 289.
† The derivation of this equation will be given in Sec. 2-7.

has been placed next to it. Since the lattice is less stiff in the direction of the vacancy, ν_v will be less than ν_0, and thus $\Delta S_v > 0$. For simplicity, ΔS_v was taken proportional to ΔS_m, and the sum of the two, $\Delta S = \Delta S_m + \Delta S_v$, written

$$\Delta S \simeq \lambda \beta \frac{\Delta H}{T_m} \qquad (2\text{-}38)$$

where λ is a new constant which should be about equal to 1 and should be the same for all metals with the same crystal structure. The value of ΔS given by Eq. (2-38) agrees with the more accurate data for self-diffusion in fcc metals with $\lambda = 0.55$. For some bcc metals ΔS is positive, as Zener predicted, but in others ΔS is definitely negative. The explanation for this is still in doubt.

Some authors have tried to extend and refine the application of Zener's theory to self-diffusion in pure metals.[1] These are helpful in that they give the student a better understanding of the application of the theory. However, in attempting to use the theory to predict the mechanism of diffusion in bcc metals, they have been incorrect.[2] These errors emphasize the limitations of Eq. (2-38).

2-6. EMPIRICAL RULES FOR OBTAINING ΔH AND D_0

In spite of the availability of experimental values of ΔH and D_0 for many metals, one is often confronted with the problem of making "reasonable guesses" about the value of D in metals where no measurements are available. In these cases one must resort to empirical rules. For self-diffusion in pure metals, Zener's theory of D_0 provides one such rule. For close-packed metals it works well, although its applicability to all bcc metals is not justified (see Table 2-2).

To obtain a value of ΔH there are two rules which can be used. These are

$$\frac{\Delta H}{T_m} \simeq 36 \frac{\text{cal}}{{}^\circ\text{K}} \qquad \frac{\Delta H}{L_m} \simeq 16.5$$

where T_m is the melting point of the metal in degrees Kelvin and L_m is the heat of fusion. Some examples are given in Table 2-2. It can be seen from Table 2-2 that the rules are only approximately obeyed and should be relied upon only if there is no reasonable alternative.[3]

[1] A. LeClaire, *Acta Met.*, **1**: 438 (1953). Also F. Buffington and M. Cohen, *Acta Met.*, **2**: 660 (1954).

[2] P. G. Shewmon, *Acta Met.*, **3**: 452 (1955).

[3] For a more general empirical equation, see O. Sherby and M. Simnad, *Trans. ASM*, **54**: 227 (1961).

For alloys the fit is even worse, as will be seen below. For hexagonal metals there are two different diffusion coefficients, depending on the direction in the lattice, although again the rules given are a satisfactory first approximation. Magnesium is exceptional in that D is almost isotropic. The data on bcc metals are shown side by side to indicate the difficulty of applying Zener's theory of D_0 to some of the bcc metals.

TABLE 2-2. Comparison of Empirical Rules and Self-diffusion Data for Pure Metals

Metal	Cryst. struct.	$\Delta H \left(\dfrac{\text{kcal}}{\text{mole}}\right)$	$D_0 \left(\dfrac{\text{cm}^2}{\text{sec}}\right)$	$\Delta H/T_m \left(\dfrac{\text{cal}}{°\text{K}}\right)$	$\Delta H/L_m$	λ
Cu[1]	fcc	47.1	0.20	34.8	15.3	0.51
Ag[2]	fcc	44.1	0.40	35.8	16.3	0.49
Ni[3]	fcc	66.8	1.30	38.6	15.8	0.28
Au[4]	fcc	41.7	0.091	31.2	13.8	0.56
Pb[5]	fcc	24.2	0.28	40.4	20.4	0.41
Mg[6]	hcp	32.2 / 32.5	1.0 / 1.5	34.8 / 35.1	14.5 / 14.7	0.80
α-Fe[7]	bcc	67.2	118	37.2	18.5	0.85
γ-U[8]	bcc	27.5	0.0018	19.5	11.3	-0.46^a
β-Zr[9]	bcc	38	0.0024	18	....	-0.37^a
Nb[10]	bcc	105	12	38.4	16.4	0.63^a
Na[11]	bcc	10.5	0.24	28.3	16.5	
Ge[12]	dia.	68.5	7.8	55.6	9.3	0.73

[1] A. Kuper, H. Letaw, L. Slifkin, and C. Tomizuka, *Phys. Rev.*, **98**: 1870 (1955).
[2] C. Tomizuka and E. Sonder, *Phys. Rev.*, **103**: 1182 (1956).
[3] R. Hoffman, F. Pickus, and R. Ward, *Trans. AIME*, **206**: 483 (1956).
[4] S. Makin, A. Rowe, and A. LeClaire, *Proc. Phys. Soc.*, **70B**: 545 (1957).
[5] N. Nachtrieb and G. Handler, *J. Chem. Phys.*, **23**: 1569 (1955).
[6] P. G. Shewmon, *Trans. AIME*, **206**: 918 (1956).
[7] R. Borg and E. Birchenall, *Trans. AIME*, **218**: 980 (1960).
[8] Y. Adda and A. Kirianenko, *Comptes rendus*, **247**: 744 (1958).
[9] V. Lyashenko, *Fiz. Met. and Metallov.*, **7**: 362 (1959).
[10] R. Resnick and L. Castleman, *Trans. AIME*, **218**: 307 (1960).
[11] N. Nachtrieb, E. Catalano, and J. A. Weil, *J. Chem. Phys.*, **20**: 1185 (1952).
[12] H. Letaw, W. Portnoy, and L. Slifkin, *Phys. Rev.*, **98**: 1536 (1955).
^{a} Assuming $\beta = 0.4$ and $\nu = 9 \times 10^{12}$ sec^{-1}.

One further crude rule of thumb which the author has found useful is that D for self-diffusion at the melting point of a metal is usually about 10^{-8} cm^2/sec. This combined with the equations of Chap. 1 gives an upper limit for what can be accomplished by self-diffusion.

2-7. CALCULATION OF ΔH AND ΔS FROM FIRST PRINCIPLES

To continue our discussion of the calculation of the diffusion coefficient, we shall outline the calculations of ΔH_v, ΔH_m, ΔS_v, and ΔS_m which have been made using the models of solid-state physics. The actual calculations are beyond the scope of this book. Nevertheless, by reviewing the models and the results, the student will obtain a feeling for the physical effects which contribute to ΔH and ΔS.

Calculation of ΔS_v and ΔS_m.[1] It is shown in most texts on statistical mechanics that the Helmholtz free energy of a crystal (F) can be represented by the equation[2]

$$\Delta F = F - F^\circ = -kT \sum_i \ln \left(1 - \exp \frac{h\nu_i}{kT} \right)^{-1} \qquad (2\text{-}39)$$

where i is summed over the frequencies of the crystal. The entropy change for some process can then be obtained from Eq. (2-39) by using the thermodynamic equation

$$\Delta S = - \left(\frac{\partial \Delta F}{\partial T} \right)_V \qquad (2\text{-}40)$$

For temperatures well above the Debye temperature, $h\nu_i \ll kT$, and these two equations give

$$\Delta S = -k \sum_i \ln \frac{h\nu_i}{kT} \qquad (2\text{-}41)$$

If the frequencies of the perfect crystal are designated ν_{i0}, and the frequencies after the introduction of a defect as ν_{if}, the entropy change when a defect is introduced is

$$\Delta S = k \sum_i \ln \frac{\nu_{i0}}{\nu_{if}} \qquad (2\text{-}42)$$

The summation in Eq. (2-42) extends over all of the vibrational modes of the crystal. Actually solving for all of these modes is too complicated a problem, so Huntington et al. simplified the equations by dividing the lattice into three regions. The first region contains only the nearest neighbors of the defect, and in this region Eq. (2-42) is used. If the defect involved is a vacancy, the force required to

---

[1] The discussion given here follows that of H. Huntington, G. Shirn, and E. Wajda, *Phys. Rev.*, **99**: 1085 (1955). This article is highly recommended.

[2] See for example G. Rushbrooke, "Statistical Mechanics," chap. 2, Oxford University Press, Fair Lawn, N.J., 1951.

slightly displace an atom into the vacancy will be less than that required for the same displacement in a perfect crystal. This means that $\nu_f < \nu_0$, so that the contribution of the atoms in this region would tend to make $\Delta S > 0$. If the defect involved is an interstitial atom, the atoms next to the defect will be pushed much closer together, so that $\nu_f > \nu_0$, and this will tend to make $\Delta S < 0$.

The second region contains the elastic stress field set up by the defect. Here elasticity theory is applied. As was pointed out in the discussion of Zener's theory of D_0, the elastic moduli decrease with increasing temperature, so this elastically strained region always makes a positive contribution to ΔS.

The third region contains the rest of the lattice and is only affected by the expansion or contraction required to give zero pressure at the surface. The contribution of this region always reduces the magnitude of the contribution of the first region but does not change its sign.

For the metal copper the values of ΔS calculated for various defects are

$$\Delta S_v \simeq 1.5R$$
$$\Delta S_v + \Delta S_m \simeq 0.9R \tag{2-43}$$
$$\Delta S \text{ (interstitial)} \simeq 0.8R$$

It is seen that in each case the effect of adding one mole of a defect is to increase the entropy of the crystal by an amount roughly equal to the gas constant R. As a comparison between these calculated values and experimental results, the value of $\Delta S_v + \Delta S_m$ for copper can be obtained from data on the diffusion coefficient of copper in copper. The experimental values of $D_0 = 0.2$ cm^2/sec,[1] $\nu = 7 \times 10^{12}$ sec^{-1}, and $a_0 = 3.61$ A give $\Delta S_v + \Delta S_m \simeq 3R$. This is considered to be good agreement since the calculated value is only approximate and the value of ν to be taken is also uncertain enough to make up the discrepancy.

Calculation of ΔH_v. While the discussion of ΔS_v dealt primarily with the vibrations of the lattice atoms around the defect, the calculation of ΔH_v deals primarily with the change in energy of the nearly free electrons.[2] The discussion given here will deal specifically with copper, although the results should be similar for silver and gold. In all of these metals the ion cores, or filled electron shells, resist interpenetration, and the atoms can be thought of as relatively incompressible spheres.

[1] A. Kuper, H. Letaw, L. Slifkin, and C. Tomizuka, *Phys. Rev.*, **98**: 1870 (1955).

[2] A readable summary of these calculations is given by W. Lomer, in B. Chalmers (ed.), "Progress in Metal Physics," vol. 8, p. 255, Pergamon Press, Inc., New York, 1959.

To establish a model for calculating ΔH_v, we take advantage of the fact that the enthalpy of the crystal depends only on the number of vacancies present and not on the mechanism by which they were produced.   For this reason the conceptual procedure used here to form a vacancy need bear no resemblance to how the vacancies are actually formed in the real crystal.   We consider the metal copper to consist of ions with a charge of $+1$, arranged in a gas of electrons.   If a neutral atom is removed from the center of the crystal and placed on a rough area of the surface, there is no change in surface area, but there is an increase in the volume.   This volume increase decreases the average energy of all of the electrons and gives an energy change of -2.8 ev[1] per vacancy.[2]   The removal of an atom from the center of the specimen to the surface leaves one atomic volume devoid of charge.

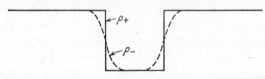

FIG. 2-14. Model for positive and negative charge density distribution (ρ_+ and ρ_-) around a vacancy.

The free electrons in the region around this vacant volume will tend to flow into the vacancy, but since there is no positive charge in the vacant site this will increase the electrostatic energy.   This can be seen with the aid of Fig. 2-14 where it is assumed that the positive charge density drops sharply at the edge of the vacant site, while the time average of the electron density tails off into the vacant site.   The greater the electron penetration of the vacancy, the greater the electrostatic energy of the separated positive and negative charge.   However, if this electrostatic energy is minimized by forming a very sharp change in the electron density, shorter wavelengths are required for the electrons and thus higher energies.   At equilibrium, the increase in energy is primarily due to the shorter wavelength (higher kinetic energy) of these electrons and is $+4.0$ ev per vacancy.

Although most of the energy change accompanying the formation of a vacancy is electronic, there is a small contribution from the change in the positions of the ions surrounding the vacancy.   These ion cores can be thought of as close-packed spheres which are slightly compressed.   The energy of interaction for each pair of nearest-neighbor

[1] ev (electron volt) is a unit of energy which is convenient for expressing the energy changes in atomic processes.   1 ev per atom = 23,060 cal/mole.

[2] The energy changes given here are those of F. Fumi, *Phil. Mag.*, **46:** 1007 (1955).

atoms can be represented by the equation

$$E(r) = A \exp\left(-\frac{r}{\rho}\right) \tag{2-44}$$

where A and ρ are empirical constants and r is the distance between the centers of the two atoms. If an atom is removed, the surrounding ions will relax into the vacancy, thereby decreasing their energy. This relaxation is small in a close-packed lattice, and the energy decrease from this source is only -0.3 ev per vacancy.

If these three contributions are added together, one obtains a value of 0.9 ev per vacancy = 21 kcal/mole = $\Delta E_v \simeq \Delta H_v$ for copper. The experimentally obtained value of ΔH_v is 1.17 ev (Table 2-3).

Calculation of ΔH_m. To calculate the energy of activation for an atom jumping into a vacancy, a model similar to Fig. 2-13b is used for an activated state. Thus an atom is placed at the saddle point, and the surrounding ions and electrons are allowed to relax to this new configuration. Consideration of the geometry shows that the vacancy has been divided into two equal halves. Thus to a first approximation there is no rearrangement of the electrons, and the electronic contribution to ΔH_m is zero. However, the lattice is distorted by the saddle-point configuration, and the ion-core interaction energy is appreciable. The calculation of this interaction for the atoms which are nearest and next nearest neighbors of the activated atom and vacancy is obtained by allowing the atoms to relax until the sum of all of the $E(r)$ terms is a minimum.[1] The value obtained for copper is 0.6 ev per atom $\simeq$ 14 kcal/mole = $\Delta E_m \simeq \Delta H_m$. The sum of $\Delta H_m + \Delta H_v$ is thus 1.5 ev per atom or 35 kcal/mole. The observed value of ΔH is 47 kcal/mole.[2] The uncertainties in the calculated ΔH are such that the difference of 12 kcal/mole is considered to be satisfactory agreement.

In summary, ΔS and ΔH can be roughly calculated, using simplified models. These calculations are not so accurate as the experimental values and can be performed only for a few metals. Nevertheless, they are accurate enough to show that the terms contributing to ΔH and ΔS are known.

One other purpose which these calculations serve is to allow comparison of the relative atomic jump frequencies by various mechanisms. Huntington and Seitz[3] first published their calculations of ΔH for copper in 1942. They obtained for pair interchange, inter-

[1] For a complete discussion of the geometry and procedure, see H. B. Huntington and F. Seitz, *Phys. Rev.*, **61**: 315 (1942).

[2] A. Kuper, H. Letaw, L. Slifkin, and C. Tomizuka, *Phys. Rev.*, **98**: 1870 (1955).

[3] H. B. Huntington and F. Seitz, *Phys. Rev.*, **61**: 315 (1942).

stitial, and vacancy mechanisms, ΔH's of ~ 200 kcal/mole, 210 kcal and 30 kcal, respectively. The vacancy mechanism was thus strongly indicated. However, this was not generally accepted, at least for alloys, until the Kirkendall experiments had been confirmed and interpreted in about 1950. (This will be discussed in Chap. 4.) It is now generally agreed that the vacancy mechanism is the predominant mechanism in close-packed metals, and these calculations remain as the most reliable indicator of the relative contribution of the three mechanisms to self-diffusion.

2-8. EXPERIMENTAL DETERMINATION OF ΔH_v, ΔH_m, AND ΔS_v

At least in principle, it is possible to perform experiments which will give values of ΔH_v, ΔH_m, ΔS_v, and ΔS_m. In practice, experiments have now been performed which, it is felt, have measured all of these quantities but ΔS_m. A few of these experiments will be discussed in the following sections. Such a discussion will establish the magnitude of these quantities for several metals, as well as establish the correctness of the theoretical calculations outlined in the preceding sections.

A variety of experiments has been performed which depend on the formation and movement of point defects. These range from studies which use a knowledge of defects and diffusion to infer mechanisms for complicated kinetic processes, to studies which are simple enough to provide unambiguous information on the concentration and diffusion of vacancies. We shall discuss two types of experiments from this latter group of experiments.[1] The first uses thermal expansion data to obtain values of ΔS_v and ΔH_v. The second quenches in vacancies and studies the rate of their annealing to give values of ΔH_v and ΔH_m.

Thermal Expansion. It was shown earlier that when a metal is heated, the equilibrium concentration of point defects increases. The type of defect which is most important in diffusion studies in metals is thought to be the vacancy. To obtain the desired values of ΔH_v and ΔS_v, it is necessary to determine the absolute value of N_v and its variation with temperature. Of the procedures used to determine ΔH_v the simplest to interpret and the only ones giving data on ΔS_v are experiments which study the thermal expansion. When a piece

---

[1] For a detailed discussion of these studies, see Lomer, *loc. cit.*, and also "Vacancies and Other Point Defects in Metals and Alloys," Institute of Metals Monograph 23, London, 1958. A closely related controversy, which will not be discussed here, concerns the effect of creep on diffusion. See Darby, Tomizuka, and Balluffi, *J. Appl. Phys.*, **32**: 840 (1961), and C. Lee and R. Maddin, *J. Appl. Phys.*, **32**: 1846 (1961).

of metal is heated, its length L increases. This expansion stems partly from an increase in the distance between lattice planes, but also from the creation of additional vacant sites inside the crystal. The lattice parameter a_0, as measured with X rays, also increases with temperature; however, the a_0 determined by X rays measures only the increase in the average distance between lattice planes. Thus the increase in the atom fraction of sites $\Delta n/n$ will be proportional to the difference between the increase in length of a sample $\Delta L/L$, and the increase in lattice parameter $\Delta a/a$. A more detailed analysis shows that for a cubic metal the relation is[1]

$$\frac{\Delta n}{n} = 3\left(\frac{\Delta L}{L} - \frac{\Delta a}{a}\right) \qquad (2\text{-}45)$$

The factor of 3 enters since $\Delta n/n$ is proportional to the change in volume, while $\Delta L/L$ and $\Delta a/a$ refer to changes in length. Equation (2-45) gives the increase in the fraction of atom sites independent of (1) the type of defects (vacancies or interstitials), (2) the degree of lattice relaxation around the sites, or (3) any pairing or clustering of the defects. If both vacancies and interstitials were formed, $\Delta n/n$ would be proportional to the difference between the concentrations of the two defects. In metals, $\Delta n/n$ is positive, as we would expect, and it will be assumed that $\Delta n/n$ is due entirely to vacancies.

Equation (2-45) is quite simple, but the experimental measurements required to use it are far from simple. Near the melting point $\Delta n/n \simeq 10^{-4}$. Thus, to measure $\Delta n/n$ with an accuracy of only 10% requires that $\Delta a/a$ and $\Delta L/L$ be measured to within one part in 10^5. This is a nontrivial task at room temperature, and at 700 to 1000°C it becomes a major undertaking. To minimize the effect of errors in temperature measurement, it is best to measure $\Delta a/a$ and $\Delta L/L$ on the same specimen at the same time. Careful studies of this type have been reported by Simmons and Balluffi.[2] Their results for aluminum are shown in Fig. 2-15. The difference between the two curves gives the following equation:

$$\frac{\Delta n}{n} = \exp 2.4 \exp\left(-0.76\ kT\right) \qquad (2\text{-}46)$$

At the melting point of aluminum this gives $\Delta n/n = 9.4 \times 10^{-4}$. The fact that $\Delta n/n$ is positive confirms the belief that vacancies are the dominant type of defect. If there is no interaction between vacancies,

---

[1] R. Simmons and R. Balluffi, *Phys. Rev.*, **117**: 52 (1960).
[2] R. Simmons and R. Balluffi, *Phys. Rev.*, **119**: 600 (1960).

they are randomly distributed. In this case $\Delta n/n$ would equal N_v, and Eq. (2-46) would indicate that

$$\Delta H_v = 0.76 \text{ ev per vacancy} = 17.5 \text{ kcal/mole,}\dagger \text{ and } \Delta S_v/R = 2.4$$

It is generally agreed that there is a small interaction between the vacancies so that the vacancies will not be randomly distributed. If this is true, the numbers appearing in Eq. (2-46) are not identically equal

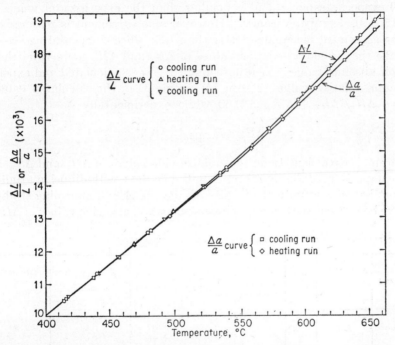

FIG. 2-15. Values of length change and lattice parameter change vs. temperature for aluminum, taking $\Delta L/L$ and $\Delta a/a$ equal to zero at 20°C. The difference between the two lines is directly proportional to the concentration of vacant atomic sites. [*From R. Simmons and R. Balluffi, Phys. Rev.*, **117**: 52 (1960).]

to $\Delta S_v/R$ and ΔH_v. This effect will be discussed more fully below, but it is felt that the changes in ΔH_v and $\Delta S_v/R$ will be $\leq 3\%$. Silver, copper, and gold have also been studied in this way, and the results are listed in Table 2-3.

† Throughout this section ΔG's and ΔH's will be quoted either as the energy change per mole of defects (kcal/mole) or as the energy per defect. This mirrors what the reader will find in the literature and should lead to no confusion since the terminology and units are completely interchangeable. The corresponding molar and atomic entropies are expressed in the dimensionless quantities $\Delta S/R$ and $\Delta S/k$, respectively.

The only metal for which a calculated value of ΔS_v exists is copper. The experimental value is $\Delta S_v/R = 1.5 \pm 0.5$, while the calculated value of Eq. (2-43) gave $\Delta S_v/R = 1.5$. The agreement is certainly satisfactory, but the reader should remember that the approximations in the calculated and experimental values are such that agreement to within a factor of 2 would also have been satisfactory.

The calculated value of ΔH_v for copper, quoted in Sec. 2-7, was 0.9 ev per vacancy. This is smaller than the experimental value of 1.17 ± 0.11 ev per vacancy. However, the agreement is reasonable. The agreement is actually better than this, since compensating errors make the calculated relative values of ΔH_v and ΔH_m more valid than their absolute values. The agreement between calculated and experimental relative values is seen by noting that the calculated values give $\Delta H_v/(\Delta H_v + \Delta H_m) \simeq 0.60$ while experimentally

$$\Delta H_v/\Delta H_{\text{self-diff.}} = 0.57$$

Thermal expansion studies do not provide values of ΔH_m and ΔS_m, so a comparison of Simmons's and Balluffi's data with diffusion studies must be delayed until the results of quenching and annealing studies have been discussed below. Those studies will also give values of ΔH_m.

TABLE 2-3. ΔH_v, ΔH_m, and ΔS_v for Some Metals

Metal	$\Delta n/n \times 10^4$ at melting temp.	$\Delta S_v/k$	ΔH_v (ev) thermal expansion	ΔH_v (ev) quenching	ΔH_m (ev) quenching	$\Delta H_v + \Delta H_n$ (ev) quenching	ΔH (ev) self-diffusion[j]
Au	7.2[a]	1.0[a]	0.94[a]	0.98[b]	0.82[b]	1.80	1.80
				0.95[c]	0.68[c]	1.63	
				0.97[d]	0.6–0.7[d]		
Ag	1.7[a]		1.09[a]	1.10[e]		1.84	1.91
				1.01[f]	0.83		
				1.06[g]			
Al	9.0[a]	2.2[a]	0.75[a]	0.79[h]	0.52[h]	1.31	
				0.76[i]	0.44[i]	1.20	
Cu	2.0[k]	1.5[k]	1.17[k]				2.04

[a] R. Simmons and R. Balluffi, *Phys. Rev.*, vol. 125 (1962).

[b] J. Bauerle and J. Koehler, *Phys. Rev.*, **107**: 1493 (1957).

[c] F. Bradshaw and S. Pearson, *Phil. Mag.*, **2**: 379 (1957).

[d] W. DeSorbo, *Phys. Rev.*, **117**: 444 (1960).   (Microcalorimetric technique.)

[e] M. Doyana and J. Koehler, *Phys. Rev.*, **119**: 939 (1960).

[f] S. Gertsriken and N. Novikov, *Phys. Metals and Metallog.*, 9(2): 54 (1960).   (Thermoelectric technique.)

[g] Y. Quere, *Compt. Rend.*, **251**: 367 (1960).

[h] W. DeSorbo and D. Turnbull, *Acta Met.*, **7**: 83 (1957).

[i] F. Bradshaw and S. Pearson, *Phil. Mag.*, **2**: 570 (1957).

[j] See Table 2-2.

[k] R. Simmons and R. Balluffi, private communication.

Quenching Experiments. If a metal is heated, the new, higher equilibrium concentration of vacancies is established first at dislocation and boundaries, which act as sources. The new concentration is then spread throughout the specimen by the diffusion of vacancies out into the crystal. If the specimen is cooled, the sources act as sinks, and the vacancy concentration of the sample is lowered by the diffusion of vacancies to these sinks. In either case, a finite time is required to reach the new equilibrium vacancy concentration. If a metal is cooled very rapidly to a low temperature, most of the vacancies do not have time to diffuse to sinks and are said to be "quenched in."

The electrical resistance provides a sensitive measure of the vacancy concentration, so that under special conditions it can be used to measure the number of vacancies quenched in and the rate at which they anneal out. The specific resistance of a pure metal ρ can be thought of as being made up of two parts—one part due to the thermal oscillations of the lattice $\rho(T)$ and a second part due to various defects in the lattice such as vacancies, impurity atoms, and dislocations. Since we wish to consider the vacancies separately, we can represent this latter term by $\rho_d + \rho_v$, where ρ_d is all defects aside from vacancies. The equation for ρ is thus

$$\rho = \rho(T) + \rho_d + \rho_v \tag{2-47}$$

To measure the resistance change due to some quenching or annealing operation, it is necessary to measure ρ at the same low temperature before and after the operation. This makes $\rho(T)$ the same in all measurements and also makes it small. To make ρ reflect only changes in ρ_v, it is also necessary to keep ρ_d from changing during the cycle. This will be accomplished if the specimen is not contaminated on heating or quenching and if no additional dislocations are introduced by quenching stresses. At the low vacancy concentrations involved, ρ_v will be proportional to N_v, so that if $\rho_d + \rho(T)$ is the same before and after a cycle, we shall have

$$\Delta\rho = \Delta\rho_v = \alpha \, \Delta N_v \tag{2-48}$$

Using a resistance bridge, the electrical resistance of a metal specimen can be measured with great precision. Experimentally this means that if a pure metal is held at the temperature of liquid nitrogen [so $\rho(T)$ is small], a change in vacancy concentration will show up as an easily and accurately measurable change in ρ. Though the actual measurements involved are easier to make than those required in the thermal expansion experiments, the interpretation is plagued by such questions as: was the specimen deformed on quenching; was it contaminated, what fraction of the vacancies was quenched in, etc.?

If the assumptions are met, two types of information can be obtained from quenching experiments.[1] The first concerns the temperature dependence of N_v. As the specimen is quenched from higher temperatures, a larger number of vacancies will be quenched in. If all of the vacancies, or even a constant fraction, are quenched in, ΔH_v can be obtained from the variation of the quenched-in resistance $\Delta\rho$ with the quenching temperature T_q. The quenched-in concentration of vacancies will be orders of magnitude larger than the initial concentration, so that Eq. (2-48) can be rewritten

$$\frac{\Delta\rho}{\alpha} = \Delta N_v = N_v = \exp\left(-\frac{\Delta G_v}{RT_q}\right)$$

or
$$\Delta\rho = A \exp\left(-\frac{\Delta H_v}{RT_q}\right) \tag{2-49}$$

Several groups have performed experiments in which the quenched-in resistance was found to obey this equation. The resulting values of ΔH_v from a few of the studies are listed in Table 2-3 along with the values determined from thermal expansion. The agreement between the values of ΔH_v obtained by the two techniques is within experimental error.

The second type of information which is obtained from quenching experiments comes from a study of the rate at which the supersaturation of vacancies anneals out. If the supersaturation is removed by the diffusion of vacancies to fixed sinks, the change in the average vacancy concentration can be approximated by the first term of an infinite series, e.g., see Eq. (1-40). Thus

$$\alpha\,\Delta N_v(t) = \Delta\rho(t) \simeq \Delta\rho_0 \exp\left(-\frac{t}{\tau}\right) \tag{2-50}$$

This equation is found to be obeyed if the concentration of quenched-in vacancies is not too high, that is, if T_q is not too high.[2] The experimental procedure is to measure the initial $\Delta\rho$ of the quenched wire, anneal at a higher temperature ($\simeq 40°C$ for gold) for several hours, and then cool to a low temperature again, to measure the change in ρ. The relaxation time τ is then obtained from a plot of $\ln(\Delta\rho/\Delta\rho_0)$ versus time.

[1] How well the assumptions are fulfilled is discussed in detail in the original papers on the subject. The interested reader is referred to the appropriate references listed in Table 2-3.

[2] For example, see J. Bauerle and J. Koehler, *Phys. Rev.*, **107**: 1493 (1957).

The relaxation time τ is given by an equation of the form $\tau = l^2/\beta D_v$ when l is the distance between sinks, β is a constant which depends on the sink geometry, and D_v is the diffusion coefficient for the vacancies. Now $D_v = \frac{1}{6}\Gamma_v a^2 = a_0^2 w$ and can be expressed as

$$D_v = B \exp\left(-\frac{\Delta H_m}{RT}\right) \tag{2-51}$$

where B is a constant. The sink geometry is too complicated to allow the evaluation of D_v from τ, but for a given quenched specimen the

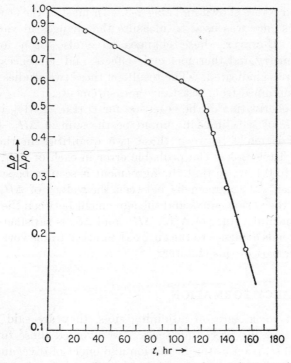

FIG. 2-16. Variation of annealing rate of quenched-in resistance with annealing temperature. This specimen was quenched from 700°C, annealed at 40°C for 120 hr and then at 60°C. From the ratio of the two slopes, ΔH_m is calculated to be 0.82 ev. [*From J. Bauerle and J. Koehler, Phys. Rev.*, **107**: 1493 (1957).]

sink geometry will be time-independent. If this is true, ΔH_m can be evaluated from the ratio of the values of τ for two successive annealing temperatures, that is,

$$\frac{\tau_1}{\tau_2} = \frac{D_v(T_2)}{D_v(T_1)} = \exp\left[-\frac{\Delta H_m}{R}\left(\frac{1}{T_2} - \frac{1}{T_1}\right)\right] \tag{2-52}$$

Figure 2-16 shows a semilog plot of $\Delta\rho(t)/\Delta\rho(0)$ versus t in which the annealing temperature was changed from 40 to 60°C after the annealing out of about 50% of the excess vacancies. The value of ΔH_m obtained therefrom is 0.82 ev per vacancy (18.9 kcal/mole).

It should be emphasized that the first equality in Eq. (2-52) holds only if the factor l^2/β is the same for both temperatures. This will be true for temperature changes at low temperature. However, if the specimen is reheated and quenched again, new sinks may be formed and l^2/β may be different.

The values of ΔH_m found in some of the more careful studies on Ag, Au, and Al are shown in Table 2-3. In most of the studies listed, electric resistance was used to measure the changes in vacancy concentration. However, these changes have also been followed by microcalorimetry and thermoelectric effects, and references d and f to Table 2-3 are so indicated. The results of these two studies agree with the values obtained from resistance measurements.

From the derivation of the equation for D [Eq. (2-31)], it is known that the ΔH of self-diffusion should be the sum of $\Delta H_m + \Delta H_v$. A comparison is made between these two quantities in the last two columns of Table 2-3. The probable error in each of ΔH_v, ΔH_m, and ΔH is 0.05 to 0.1 ev, so that the agreement is seen to be satisfactory.

In summary, the agreement between the values of ΔH_v and ΔH_m obtained in the various ways and the agreement between the calculated and experimental values of ΔH_m, ΔH_v, and ΔS_v is satisfactory. This agreement lends credence to the entire theoretical framework developed in the latter part of this chapter.

2-9. DIVACANCY FORMATION

If two vacancies are on adjoining sites, they are said to form a divacancy. In the preceding section we avoided those topics which required a discussion of the formation and effect of divacancies. The results given above are correct, but a satisfactory understanding and interpretation of the experiments require a better understanding of divacancies. The development given here is primarily a thermodynamic one. It is thus also applicable to the related problems of solute-pair formation and the formation of vacancy-solute pairs in crystals.

Consider the reaction of two vacancies to form a divacancy. If there is no interaction between the two, the atom fraction of such pairs (N_{v2}) in an fcc lattice is given by the equation

$$N_{v2} = 6N_v{}^2 \tag{2-53}$$

This is obtained as follows. The number of vacancies that are in divacancies in a mole of material is given by the number of vacancies n_v times the probability that there will be a vacancy on any one of the vacancy's twelve nearest-neighbor sites p_v. In the absence of any interaction between vacancies, the probability that there is a vacancy on any specific site is N_v. Therefore $p_v = 12N_v$, and the number of vacancies in divacancies is $12n_vN_v$. This means that the number of divacancies is $6n_vN_v$, and if we define N_{v2} as this number over the number of sites in a mole, Eq. (2-53) is obtained.

If there is an interaction between two vacancies, Eq. (2-53) will not be valid. The more interesting situation is the one in which the two vacancies are attracted to one another. In this case, the free energy of the lattice is lowered when two vacancies form a divacancy, or equivalently, when work is required to separate two adjacent vacancies. As a result N_{v2} will be greater than $6N_v^2$. The equation for N_{v2} can most simply be obtained by forming the equilibrium constant for the reaction. If we consider the formation of divacancies of a particular orientation, the equation can be schematically written

$$\square + \square \leftrightharpoons \square\square$$

and, from the law of mass action[1]

$$\ln \frac{N_{v2}}{N_v^2} = - \frac{\Delta G_2}{RT} \quad \text{or} \quad N_{v2} = N_v^2 \exp\left(- \frac{\Delta G_2}{RT}\right) \quad (2\text{-}54)$$

However, there are six different orientations of divacancies in an fcc lattice, and we wish to count all of them in N_{v2}. Therefore

$$N_{v2} = 6N_v^2 \exp\left(- \frac{\Delta G_2}{RT}\right) \quad (2\text{-}55)$$

Here ΔG_2 is the molar free-energy change of the lattice when divacancies are formed from separated vacancies.[2]

Divacancy formation entered the interpretation of the thermal expansion studies only in relation of $\Delta n/n$ to N_v. This comes about from the fact that

$$\frac{\Delta n}{n} = N_{v1} + 2N_{v2} + \cdots + nN_{vn} + \cdots \quad (2\text{-}56)$$

[1] The reader who is not familiar with this law will find it discussed under this title or "equilibrium constant" in most books on thermodynamics or physical chemistry, e.g., L. Darken and R. Gurry, "Physical Chemistry of Metals," chap. 9, McGraw-Hill Book Company, Inc., New York, 1953.

[2] An additional check on the result can be obtained by noting that if $\Delta G_2 = 0$ (no interaction), Eqs. (2-55) and (2-53) give identical results.

where N_{v1} is the mole fraction of vacancies present as individual vacancies.[1] Considering only the first two terms in Eq. (2-56), it can be rewritten

$$\frac{\Delta n}{n} = N_{v1}\left[1 + 12N_{v1} \exp\left(-\frac{\Delta G_2}{RT} \right) \right] \qquad (2\text{-}57)$$

The term ΔG_2 has yet to be accurately measured, but there are several experiments which indicate that it is roughly -0.1 to -0.2 ev. The effect of such a value of ΔG_2 is to make ΔH_v slightly less than Q, where Q is defined by the equation

$$\frac{\Delta n}{n} = \alpha \exp\left(-\frac{Q}{RT} \right)$$

The effect is certainly small, and until ΔG_2 is known, it is satisfactory to ignore it.

The effect of divacancies on the quenching experiments can be much more pronounced. This stems from two results: first, as the temperature decreases the fraction of vacancies in divacancies, at equilibrium, becomes much larger; second, at the annealing temperatures used in the quenching experiment, atoms will jump into a divacancy with a higher frequency than into an individual vacancy. These effects do not appear to affect the value of ΔH_v obtained from the quenched specimens, but they completely change the annealing kinetics of the vacancies. Bauerle and Koehler found in gold that if the quenching temperature was greater than 700°C, the excess resistance (vacancies) did not anneal out according to Eq. (2-50), but annealed out more rapidly.[2] At the higher quenching temperatures more vacancies are formed. During annealing many of these combine with each other before they get to dislocations. These divacancies then diffuse more rapidly to the dislocation. The detailed analysis is difficult and uncertain;[3] we only wish to show here why the analysis leading to Eq. (2-50) is no longer valid. It is of interest though that the detailed analysis indicates $\Delta G_2 \simeq -0.2$ ev per pair and that the enthalpy of activation for the movement of an atom into an adjacent divacancy is about 0.2 ev less than ΔH_m.

---

[1] Prior to this we have used N_v to represent the number of vacancies divided by the number of sites, independent of where the vacancies are relative to one another. In this section it is desirable to distinguish between N_v and N_{v1}. However, $N_{v1} \approx N_v(1 - 12N_v)$ and since $N_v \leq 5 \times 10^{-4}$, it is normally satisfactory to take $N_{v1} = N_v$.

[2] J. Bauerle and J. Koehler, *Phys. Rev.*, **107**: 1493 (1957).

[3] F. Seitz, J. Koehler, and J. Bauerle, *Phys. Rev.*, **107**: 1499 (1957).

2-10. EFFECT OF HYDROSTATIC PRESSURE ON DIFFUSION

Over the last decade the study of the effect of pressure on diffusion has developed into a valuable, though specialized, tool. As an example of what can be learned from such studies, we shall consider the case of self-diffusion in pure metal by a vacancy mechanism.

In Sec. 2-4 it was shown that the equation for D in this case is

$$D = a^2 \nu N_v w$$
$$= a^2 \nu \exp \frac{-\Delta G_v}{RT} \exp \frac{-\Delta G_m}{RT} \tag{2-58}$$

Differentiating with respect to pressure

$$\left[\frac{\partial \ln (D/a^2 \nu)}{\partial P} \right]_T = -\frac{1}{RT} \left[\left(\frac{\partial \Delta G_v}{\partial P} \right)_T + \left(\frac{\partial \Delta G_m}{\partial P} \right)_T \right] \tag{2-59}$$

From thermodynamics we have the relation

$$\left(\frac{\partial \Delta G}{\partial P} \right)_T = \Delta V \tag{2-60}$$

If the assumptions of the derivation of the equation for w are valid (see Sec. 2-4), ΔG_m is a free energy, as is ΔG_v, and we can use Eq. (2-60) to give

$$\left[\frac{\partial \ln (D/a^2 \nu)}{\partial P} \right]_T = -\frac{1}{RT} \left(\Delta V_v + \Delta V_m \right) \equiv -\frac{\Delta V_a}{RT} \tag{2-61}$$

where ΔV_a is called the activation volume.

ΔV_v is the partial molar volume of the vacancies. Its magnitude will depend on the degree to which the atoms surrounding a vacancy relax into it. If there was no such relaxation, ΔV_v would equal the molar volume of the metal ($\bar{V}$) since creating a mole of vacancies in the specimen would increase the volume of the large piece just as much as adding a mole of the pure metal would. In any metal there will be some relaxation though, so ΔV_v will be less than $\bar{V}$. For a close-packed lattice, ΔV_v should be an appreciable fraction of $\bar{V}$, while for a bcc lattice it would be a smaller fraction of $\bar{V}$.

Physically, the removal of vacancies decreases the volume of the specimen. Equation (2-61) simply reflects the fact that if the pressure is increased, the specimen will lose vacancies in an effort to relieve the pressure increase. This decrease in the concentration of vacancies will in turn decrease D.

The second term contributing to the activation volume, that is, ΔV_m, is the partial molar volume of activated complexes. At the saddle point the diffusing atom is expanding a constriction, while the volume of the divided vacancy is approximately unchanged. Thus, ΔV_m is probably positive, but small. This means that an increase in the pressure would decrease the concentration of activated complexes, and this would also decrease D.

To show the changes in D which are involved, Fig. 2-17 shows a plot of log D versus pressure, for lead. The activation volume of

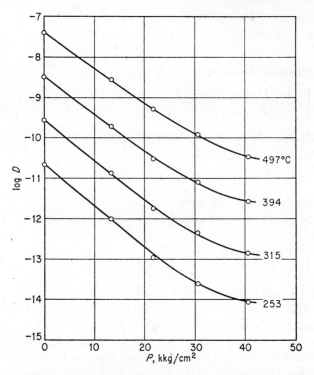

FIG. 2-17. Variation of the self-diffusion coefficient of lead with pressure. [*After J. Hudson and R. Hoffman, Trans. AIME*, **221**: 761 (1961).]

11.6 cm³/mole corresponds to a decrease in D of about a factor of 10 with each increase in pressure of 10^4 kg/cm².†

Table 2-4 gives some experimental values of ΔV_a and $\bar{V}$. The value of $\Delta V_a/\bar{V}$ for fcc lead is seen to be larger than that for the less close-packed sodium lattice. In both cases, though, the ratio is less than unity. This indicates that there is an appreciable relaxation of the

† Hudson and Hoffman point out that the apparent decrease in activation volume between 20 and 40 kkg/cm³ might be due to an error in the calibration of the pressure and base their discussion on the straight line between 0 and 20 kkg/cm³.

atoms into a vacancy. The individual values of ΔV_v and ΔV_m have not been measured separately for lead, so that the observed value of ΔV_a cannot be broken up into these two terms. However, an idea of the relative values of these terms can be obtained from other studies. Experiments on the effect of pressure on the rate at which quenched-in vacancies anneal out show that increasing the pressure decreases the rate of annealing. In gold, $\Delta V_m = 0.15\bar{V}$.[†] Using a model similar to that used in the calculation of ΔH_m and ΔH_v, Tewordt[1] has calculated $\Delta V_v \simeq 0.55\bar{V}$ for copper. The sum of ΔV_v and ΔV_m for these two close-packed metals agrees with the observed value for lead of $0.64\bar{V}$, thus the results of the various studies are consistent.

TABLE 2-4. Activation Volumes for Solid and Liquid Diffusion

Metal	$\bar{V}$ (cm³/mole)	ΔV_a (cm³/mole)	$\Delta V_a/\bar{V}$	Ref.
Na (s)	23.0	12.4	0.54	2
Pb (s)	18.25	11.6	0.64	1
Hg (l)	14.85	0.59	0.040	2
Ga (l)	11.46	0.55	0.048	2

[1] J. Hudson and R. Hoffman, *Trans. AIME*, **221:** 761 (1961).

[2] N. Nachtrieb, in "Liquid Metals and Solidification," p. 49, ASM, Cleveland, 1958.

It has been proposed that diffusion in a liquid occurs when a hole opens up in the nearest-neighbor shell of atoms, and the diffusing atom jumps into it. Some people conceive of these holes as having a volume comparable to that of a vacancy in a crystal. The effect of pressure on D is an ideal way to measure the mean size of these holes. The measured values of ΔV_a indicate that, at least in mercury and gallium, these holes are such a small fraction of $\bar{V}$ that the concept can be more misleading than helpful. A more complete discussion of this has been given by Nachtrieb.[2]

PROBLEMS

2-1. Assume that in pure copper an interstitial copper atom rests at the middle of the cube edge. In a sequence of drawings of a (100) plane show the atom movements involved in diffusion by an interstitialcy mechanism.

[†] R. Emrick, *Phys. Rev.*, **122:** 1720 (1961).
[1] L. Tewordt, *Phys. Rev.*, **103:** 103 (1958).
[2] N. Nachtrieb, in "Liquid Metals and Solidification," p. 49, ASM, Cleveland, 1958.

2-2. If at $t = 0$, a quantity of solute is located at the point $r = 0$ in a three-dimensional medium, the concentration of solute at any point r from the origin, after time t, is

$$c(r,t) = \frac{\gamma}{t^{\frac{3}{2}}} \exp\left(-\frac{r^2}{4Dt}\right)$$

(a) Give the probability (normalized to 1) of finding an atom in a spherical shell dr thick and r from the origin.

(b) What is the mean square value of r, that is, $\overline{r^2}$, for the solute after time t?

(c) Using the results of part (b) and the random-walk equation $\overline{r^2} = n\alpha^2$, show that

$$D = \tfrac{1}{6}\Gamma\alpha^2$$

where $\Gamma = n/t$.

2-3. (a) Calculate γ for a tracer in a pure bcc metal where γ is defined by the equation

$$D = \gamma a_0^2 w N_v$$

(b) Calculate γ for an interstitial solute in a dilute bcc binary alloy. (Figure 3-1 shows the interstitial sites for a bcc lattice.)

2-4. One hundred jumping beans are placed along the center line of a gymnasium floor at 6-in. intervals. Twelve hours later the distance of each from the line is measured, and the sum of the squares of the distance divided by 100 is 36 in.[2]

(a) Calculate the diffusion coefficient of the jumping beans.

(b) If the mean jump (or roll) distance of a bean is equal to 0.1 in., estimate the mean jump (or roll) frequency of a bean.

2-5. Derive an equation for the D of an interstitial solute in a binary alloy for the situation in which $w_{12} - w_{21} = (a_0/2)(\partial w/\partial c)(\partial c/\partial x) \neq 0$. [Use a procedure similar to that leading to Eq. (2-17).]

2-6. In H_2 at 1 atm and 25°C, the average molecular velocity is 13×10^4 cm/sec and the mean free path is 19×10^{-6} cm. Calculate the diffusion coefficient of the gas. (Take the average velocity to be the same as the root-mean-square velocity.)

2-7. Calculate the ratio of the equilibrium atom fractions of interstitial atoms and vacancies in copper at 1000°C. Take the ΔH to form copper interstitial atoms to be 210 kcal/mole and the ΔH to form vacancies as 30 kcal/mole. Assume that the entropy of formation is the same for both defects. (The numbers used here for ΔH are those given by Huntington in "Atom Movements," p. 74, ASM, Cleveland, 1951.)

2-8. In the temperature range −70 to 400°C the diffusion coefficient for carbon in α-iron is $D = 0.020 \exp(-20{,}100/RT)$ cm²/sec. If the average vibration frequency of a carbon atom in the lattice is 10^{12} sec^{-1}, calculate the quantity ΔS_m for carbon diffusion.

2-9. (a) Using the data given below, make a plot of log D versus $1/T$, and estimate, by eye, the best straight line through the points.

(b) Calculate ΔH and D_0 for this line.

(c) Calculate ΔH and D_0 using a least-squares procedure, assuming all error to be in the values of D. Plot the least-squares line on the graph of part (a).

D (cm²/sec)	10^{-8}	10^{-9}	10^{-10}	10^{-11}
T(°K)	1350	1100	950	800

2-10. Using the empirical rules cited in the text, give an equation for the self-diffusion coefficient of Pt.

2-11. Calculate the equilibrium concentration of triangular trivacancies in an fcc lattice (as opposed to linear trivacancies) if the free-energy decrease on combining a divacancy and a vacancy is ΔG_3.

(*a*) When $\Delta G_3 = 0$.

(*b*) When $\Delta G_3 \neq 0$.

2-12. Concerning the annealing out of excess vacancies in gold:

(*a*) If $\Delta H_m = 0.82$ ev for gold, estimate D_v at 40°C.

(*b*) If $\tau = l^2/\pi^2 D_v$, estimate the mean intersink distance l if $\tau = 200$ hr at 40°C.

(*c*) The activation energy to move an atom into a divacancy (ΔH_{m2}) is given by $\Delta H_{m2} = \Delta H_m - 0.2 = 0.62$ ev. Assuming that D_0 is the same for D_v and D_{v2} (divacancies), calculate D_v/D_{v2} at 40°C.

2-13. In the pure metal carnegium the dominant diffusion mechanism is thought to be an interstitialcy mechanism. A self-diffusion experiment shows that at 1000° K a pressure of 10^4 kg/cm^2 increases D by a factor of 8.

(*a*) Is the experimental result qualitatively consistent with an interstitialcy mechanism?

(*b*) Calculate ΔV_a for carnegium.

2-14. Would you expect ΔV_a to be positive or negative for a ring mechanism? Explain.

chapter 3 DIFFUSION IN
DILUTE ALLOYS

The next degree of complexity after studying diffusion in pure metals is to study diffusion in dilute alloys. The simplest problem in this area arises in interstitial alloys. Here the solute atoms diffuse on a sublattice whose sites are essentially all vacant, and the only role played by the solvent atoms is to form the barriers which define the sublattice of the interstitial sites. Because the two types of atoms do not share the same sites, the theory of interstitial diffusion is relatively simple and has been discussed in Chap. 2. In this chapter the discussion will be restricted to the use of relaxation or resonance techniques to measure D for interstitials in bcc metals. This technique is representative of a family of techniques in which the mean jump frequency of the interstitials is obtained from some relaxation phenomenon. This frequency is then combined with a model and random-walk theory to give values of D.

The problem of an atomic theory of D for a substitutional solute is different. Here the solute and the solvent atoms share the same sites, and the problem of analyzing D becomes more complex. Since the two atoms do share the same sites, the difference in D for the solute and the solvent atoms allows one to estimate the ratio of the jump frequencies for the solute and the solvent atoms. There are several theories which suggest why, and by how much, these D's should differ. These constitute the second section of the chapter.

The final section discusses the correlation between the directions of successive atomic jumps.   If an atom diffuses by a vacancy mechanism, the successive jump directions of the atom will not be random since the probability that the atom will jump back into the vacancy it last exchanged with is greater than the probability of its jumping to any other neighboring site.   The inequality of these probabilities gives rise to a correlation between the directions of successive jumps. In pure metals the error caused by ignoring this is small.   However, if vacancies and solute atoms tend to form pairs, this correlation becomes marked, and new equations relating the jump frequency and the vacancy concentration must be derived.   This analysis can become quite complicated and is still being developed.   However, the basic outline is well established and will be discussed.

3-1. ANELASTICITY DUE TO DIFFUSION

In a bcc lattice such as α-iron, an interstitial atom such as carbon strains the lattice more along one of the cubic directions of the lattice than along the other two.   If the interstitial jumps to a neighboring site, the direction of this high strain changes.   If a stress is applied to the crystal, the resulting strain interacts with the strain set up by the interstitials to make some of them jump into sites which align their strain fields with that of the applied strain.   This alignment gives rise to an additional strain called the anelastic strain.   From the rate at which this anelastic strain appears, the jump frequency can be determined and from this the diffusion coefficient.

To give a clearer picture of the distortion associated with an interstitial solute, consider the interstitial atom shown as a large black circle in the bcc lattice of Fig. 3-1.   Its two nearest neighbors are shown as $\otimes$, and their normal sites are a distance $a_0/2$ from the center of the solute.   The four solvent atoms which are next nearest neighbors to the interstitial (labeled e, f, g, and h) lie in the xy plane; their centers are $a_0/\sqrt{2} = 0.71a_0$ from the center of the solute atom.   If now the matrix atoms in Fig. 3-1a are enlarged until they touch one another, as is a reasonable model for a transition metal, it is seen that the distortion caused by the interstitial atom shown will be much more severe in the z direction than in either the x or the y direction.   Thus the strain field introduced by the solute is said to have tetragonal symmetry.

If the interstitial now jumps to the interstitial site to its left (shown by a small black circle in Fig. 3-1), its surroundings will be equivalent, but the tetragonal axis of the distortion will be in the y direction. If the site occupied by the interstitial in Fig. 3-1 is called a z site, consideration of Figs. 3-1*a* and 3-1*b* will show that atoms on *z* sites have only *x* and *y* sites as nearest neighbors; similarly, *x* sites have only *y* and *z* neighbors, etc.

Consider now a bcc single-crystal wire with the [001] direction along its axis and its interstitial atoms uniformly distributed between *x*, *y*, and *z* sites. If a small weight is hung on the wire, there will be an immediate elastic strain ϵ_e. As time passes, additional strain appears, though the strain rate continually decreases.[1] This latter strain is due to a net flux of atoms jumping into the *z* sites from *x* and *y* sites

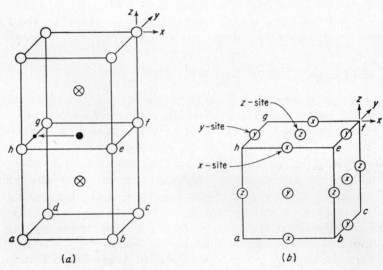

(a)   (b)

Fig. 3-1. (*a*) Body-centered cubic lattice showing interstitial atom (•), its two nearest neighbors (⊗), and an arrow pointing to one of the four sites the interstitial can jump into. (*b*) shows the interstitial sites in a bcc unit cell. There are no sites inside the unit cell so all of the types of sites are shown on the visible surfaces of the cube.

and is called the anelastic strain ϵ_a (see Fig. 3-2). This preference for *z* sites comes from the fact that the *z* sites are slightly longer in the *z* direction. Thus the strain energy associated with an occupied *z* site is less than for the *x* or *y* sites. If the load is removed, there will again be an immediate elastic strain $-\epsilon_e$ and then, since all types of sites will now be equivalent, there will be a slow, anelastic[2] strain $-\epsilon_a$ which restores the specimen to its original length.

[1] The weights involved are such that the total strain is the order of 10^{-5} or less.

[2] Since the strain is not a single-valued function of the stress, the wire is not elastic. However, when the stress is removed, the strain returns to zero. To differentiate this type of nonelastic behavior from the type in which a permanent set occurs, Zener has coined the word "anelastic."

Analysis of the Relaxation. Since the anelastic strain ϵ_a stems from interstitial atoms changing sites, it should be possible to relate the rate of decay of ϵ_a to the diffusion coefficient of the solute; in fact, this relation can be made very simply. We start by designating the number of interstitial atoms on x, y, and z sites as n_x, n_y, and n_z, respectively. If the mean jump frequency of an interstitial is designated Γ, the rate at which atoms are leaving x sites at any instant will be Γn_x. At zero applied stress the energy of all occupied sites is the same, so the atoms leaving one type of site will go half and half to each of the

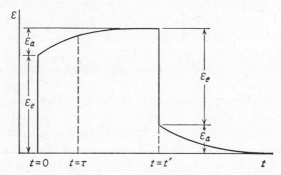

Fig. 3-2. Strain vs. time for an anelastic specimen when a load is applied at $t = 0$ and removed at $t = t'$.

other type. For example, the rate at which atoms enter x sites from y sites is $\Gamma n_y/2$. It follows that

$$\frac{dn_x}{dt} = -\Gamma n_x + \frac{\Gamma n_y}{2} + \frac{\Gamma n_z}{2} \tag{3-1}$$

but since $n_x + n_y + n_z \equiv n$ is a constant, we can replace $n_y + n_z$ to give

$$\frac{dn_x}{dt} = \frac{d(n_x - n/3)}{dt} = -\frac{3}{2}\left(n_x - \frac{n}{3}\right)\Gamma \equiv -\frac{3}{2}\Gamma\,\Delta n \tag{3-2}$$

The literature on this subject invariably talks in terms of the mean time of stay τ_j of an interstitial, which is $1/\Gamma$. Replacing $1/\Gamma$ by τ_j, Eq. (3-2) can be rewritten

$$\frac{d\,\Delta n}{dt} = -\frac{3}{2}\frac{n}{\tau_j} \equiv -\frac{n}{\tau} \tag{3-3}$$

where $(2/3)\tau_j$ has been replaced by the relaxation time τ. Integration gives

$$\Delta n = n_0 \exp\left(-\frac{t}{\tau}\right) \tag{3-4}$$

Now ϵ_a is proportional to Δn, so if log ϵ_a is plotted versus t, τ can be obtained from the slope.

The random-walk equations of Sec. 2-3 enable us to relate τ to D. Equation (2-14) gave

$$D = \tfrac{1}{6}\Gamma\alpha^2 \tag{2-14}$$

where α is the jump distance of the interstitial. Since $\alpha = a_0/2$, we have

$$D = \frac{\Gamma a_0{}^2}{24} = \frac{a_0{}^2}{36\tau} \tag{3-5}$$

To measure ϵ_a and τ experimentally, it is customary to obtain the strain by twisting a thin wire instead of pulling it in tension. The advantage of using torsion is that, by attaching a mirror to the wire and using it to cast a reflected beam of light across the room, very small strains can be easily measured. A value of τ which can be conveniently measured by this elastic aftereffect is $\tau = 0.2$ hr. This corresponds to three jumps *per hour* as compared with about 10^{12} jumps per hour for carbon in α-Fe at 730°C. The elastic aftereffect provides an accurate means of measuring values of τ from 10 to 10^4 sec. This corresponds to values of the diffusion coefficient in the range 10^{-21} to 10^{-18} cm²/sec. These values of D are orders of magnitude smaller than those that can be measured by methods based on Fick's second law, since these latter methods require measurable penetration of the solute. On the other hand, the relaxation technique reduces the required penetration to an absolute minimum: one atomic jump.

Resonance Techniques. Let us now see how this anelastic effect can be used at higher frequencies. Inspection of Fig. 3-2 shows that, after the stress is increased to some new, constant value, the elastic modulus (the ratio σ/ϵ) will be a function of time for $t \lesssim 2\tau$. If an oscillating stress is applied with a frequency $\omega \simeq 1/\tau$, this variation in the modulus, or relaxation, gives rise to a hysteresis loop when stress is plotted against strain. Equivalently, this means that as the stress is oscillated the maximum value of the strain will occur shortly after the maximum value of the stress. Mathematically this lag of the strain behind the stress by a small angle δ can be given by the equations[1]

$$\sigma = \sigma_0 \sin \omega t$$
$$\epsilon = \epsilon_0 \sin (\omega t - \delta)$$

When the stress and strain are out of phase in this manner, an energy ΔE is absorbed during each cycle. That ΔE is indeed nonzero when $\delta > 0$ can be seen by integrating over one cycle

$$\Delta E = \oint \sigma \, d\epsilon = \sigma_0 \epsilon_0 \int_0^{2\pi/\omega} \sin (\omega t) \cos (\omega t - \delta) \, dt \tag{3-6}$$

[1] In systems of interest δ usually varies between zero and 0.1 radian.

Substituting $\cos(\omega t - \delta) = \cos \omega t \cos \delta + \sin \omega t \sin \delta$ and integrating gives

$$\Delta E = \epsilon_0 \pi \sigma_0 \sin \delta \qquad (3\text{-}7)$$

Thus when $\delta = 0$, there is no energy loss, and as δ increases from zero, so does ΔE.[†]

To give a qualitative feeling for the variation of δ with frequency, we refer to Fig. 3-3. Here stress is plotted vs. strain for three ranges of $\omega\tau$.

1. $\omega\tau \gg 1$. Here the frequency is so high that essentially no interstitials can change sites in reaction to the applied stress. In this case, stress and strain are completely in phase; $\delta = 0$, and so $\Delta E = 0$.

2. $\omega\tau \simeq 1$. At this frequency, many of the interstitials will be able to change sites in reaction to the applied stress, but the stress will vary too rapidly for the equilibrium population of each type of site to be attained at any particular value of the stress. Thus $\delta > 0$, $\Delta E > 0$,

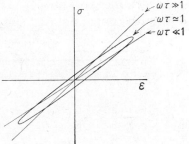

FIG. 3-3. Stress-strain curves for very small strains ($\epsilon \leq 10^{-6}$) in an anelastic material for various values of the product $\omega\tau$.

and the lag between stress and strain shows up as a hysteresis loop.

3. $\omega\tau \ll 1$. Here the frequency is so much lower than $1/\tau$ that the population of each type of site will continually be in equilibrium with the applied stress. As a result, $\delta = 0$. The only difference between this case and that of $\omega\tau \gg 1$ is a smaller slope for the stress-strain line. This stems from the fact that here the strain at any stress is the elastic plus the anelastic strain, while with $\omega\tau \gg 1$ there is no anelastic strain.

The detailed analysis of this type of phenomenon would lead us too far afield.[1] The shape of a plot of δ versus $\ln \omega\tau$ obtained from the detailed analysis is shown in Fig. 3-4 and is seen to agree with the

[†] This energy loss is expressed in several different ways in the literature. If it is observed that $(\frac{1}{2})\epsilon_0\sigma_0 = E$ is the total elastic energy in the crystal and δ is small, the most common expressions are

$$\sin \delta = \tan \delta = \delta = \frac{\Delta E}{2\pi E} = Q^{-1}$$

[1] An introduction to this is given by C. Zener, "Elasticity and Anelasticity of Metals," University of Chicago Press, Chicago, 1948. A more recent and complete review is given by A. Nowick, in B. Chalmers (ed.), "Progress in Metal Physics," vol. 4, p. 1, Pergamon Press, Inc., New York, 1953.

qualitative conclusions given above. The detailed analysis shows that the maximum in δ occurs at $\omega\tau = 1$.

In experimental studies of δ versus $\omega\tau$ there is a variety of techniques that can be used to drive the specimen, depending on the frequency needed.[1] In diffusion studies τ decreases exponentially with tempera-- ture (since $\tau = \frac{2}{3}\Gamma^{-1}$), and it is easier to vary τ continuously by changing the temperature than it is to vary ω. To determine τ, the system is oscillated at a fixed frequency, and the energy loss per cycle is meas- ured over a range of temperatures which will allow the curve of Fig. 3-4 to be traced out. From the fact that $\omega\tau = 1$ at the maximum in the

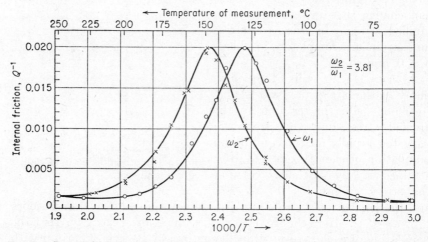

FIG. 3-4. Internal friction (δ) versus $1/T$ at two frequencies, for carbon in tantalum. Since the relaxation time τ varies exponentially with $1/T$ and ω is constant for a given curve, this is equivalent to plotting δ versus $\ln \omega\tau$. [*From T. Kê, Phys. Rev.,* **74**: 9 (1948).]

curve, τ at the temperature of the maximum is $1/\omega$. If the applied frequency is increased and δ is again measured over the same tempera- ture range, the curve will be shifted to lower temperatures, and τ can be determined at a second temperature. Once τ is known at two or more temperatures, ΔH can be calculated.

A plot of data obtained for nitrogen in niobium is shown in Fig. 3-5. The values of τ between 20 and 6,000 sec were obtained by measuring the elastic aftereffect while the values between 0.1 and 1 sec were obtained from the rate of decay of the oscillations of a torsion pendu- lum. The two techniques combine to give values of τ (and thus D)

[1] For a review of the experimental techniques, see C. Wert, "The Metallurgical Use of Anelasticity," in "Modern Research Techniques in Physical Metallurgy," p. 225, ASM, Cleveland, 1953.

ranging over five orders of magnitude. This extraordinary range of values combined with the accuracy of the individual points give values of D_0 and ΔH which are more accurate than are available with any other technique. Indeed, if it were desired, the range of the measurements could be extended still farther by going to another technique using still higher frequencies. By way of comparison, using radioactive tracers and the thin-film technique discussed in Sec. 1-3, it is difficult to work over a range of more than four orders of magnitude in D.

The applications of anelasticity are too numerous to mention here. However, the potential of the technique can be indicated by reference to one related type of study. It was mentioned in Sec. 2-9 that vacancies in gold associate to form divacancies which can diffuse faster than individual vacancies. A similar association occurs between interstitial atoms. Anelastic measurements are extremely useful in studying this latter association since they allow the measurement of the jump frequency of isolated interstitials and interstitials in pairs as well as the variation of the number of pairs with temperature. The jump frequency is determined from the frequency at which δ is a maximum. The magnitude of δ at its maximum is proportional to the number of atoms, or pairs, which give rise to the peak in question. Thus the variation of the magnitude of δ at the peak due to interstitial pairs gives the variation in the concentration of such pairs.[1]

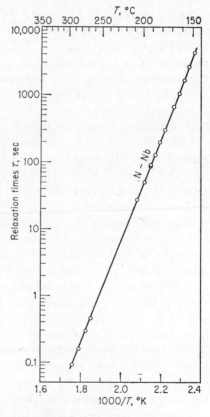

FIG. 3-5. Relaxation time versus $1/T$ for nitrogen in niobium. (*Courtesy of R. Powers.*)

Magnetic Measurement of τ. The basis for the anelastic behavior discussed above is that a stress changes the relative energy of various solute sites and the resulting ordering of the solute gives a strain. A

[1] R. Powers and M. Doyle, *Trans. AIME,* **215:** 655 (1959).

completely analogous effect appears in studying the magnetic properties of ferromagnetic bcc alloys. The initial equation for elastic phenomena was $\sigma = M\epsilon$, where M is the elastic modulus. The starting point in magnetic phenomena is $H = \dfrac{1}{\mu} B$, where μ is the permeability.

As an example, if at $t = 0$ a field H is applied to an Fe–0.01% C alloy, the resulting value of B will trace out a curve exactly analogous to that given for ϵ versus t in Fig. 3-2.[1] This lag in B stems from the rearrangement of carbon atoms. If the value of H is cycled at a frequency ω, the lag angle between B and $H(\delta)$ will vary with $\omega\tau$ as shown in Fig. 3-4. In fact, it was the temperature-dependent loss in iron transformer cores which initially led to the discovery of the role that diffusion plays in the process.

Although anelasticity is used much more frequently than ferromagnetic relaxation, the latter technique is mentioned here to emphasize the generality of relaxation effects. Still another resonance technique from which diffusion data can be obtained is nuclear magnetic resonance.[2]

Internal Friction Due to Diffusion in Substitutional Alloys. In addition to the anelastic effects in interstitial alloys discussed above, anelastic effects also appear in substitutional alloys which have been shown to result from atomic diffusion.[3] Since the peak is due to the diffusion of a substitutional element which will diffuse more slowly than an interstitial element, the relaxation peak at a given frequency will occur at a higher temperature than for interstitials. However, if an appropriate temperature range is chosen, a plot of δ versus $1/T$ is obtained which is analogous to Fig. 3-4. This effect appears on many alloys. For example, it has been studied in substitutional alloys such as zinc in fcc copper, vanadium in bcc iron, cadmium in hcp magnesium, and copper in fcc aluminum.

Zener, who originally saw this effect in copper–30% zinc, proposed that it is caused by the reorientation of isolated pairs of nearest-neighbor solute atoms. Indeed, for dilute alloys the magnitude of the peak does increase as the square of the solute concentration, as this hypothesis requires. However, this model of isolated pairs has little meaning in concentrated solutions where the effect is strongest, e.g., in alloys approaching 50% A–50% B. Other models have been

[1] G. W. Rathenau, in "Magnetic Properties of Metals and Alloys," chap. 9, ASM, Cleveland, 1959.

[2] J. Spokas and C. Slichter, *Phys. Rev.*, **113**: 1462 (1959).

[3] J. Hino, C. Tomizuka, and C. Wert, *Acta Met.*, **5**: 41 (1957). See also J. Stanley and C. Wert, *J. Appl. Phys.*, **32**: 267 (1961).

proposed,[1] but they lack the geometrical simplicity of Zener's original hypothesis. Following recent thoughts about the geometry of divacancies, deJong has attempted to describe more accurately for dilute alloys the original pair model of Zener;[2] his picture fits data for fcc alloys on the variation of the effect with crystal orientation. Clearly, the peak is due to a stress-induced ordering of some kind, but the precise mechanism for the various crystal types and compositions is unclear. Regardless of the detailed model of the effect, however, a close association exists between the kinetics of this effect and diffusion.

Precisely because of the lack of an unambiguous model, the constant γ in the equation $D = \gamma a_0^2/\tau$ cannot be evaluated from theory. Hence, the diffusion coefficient cannot be determined completely from measurement of τ. Comparisons of data from tracer measurements and from relaxation measurements on the same alloy show that γ is of the order $\frac{1}{10}$ to $\frac{1}{20}$.

Regardless of knowledge of γ, the activation energy for the relaxation should be just that of diffusion, since a_0 and γ are temperature-independent. Comparison of tracer and anelastic values of ΔH shows reasonably good agreement between the two. Recent careful experimental work on a copper–30% zinc alloy has shown that this anelastic ΔH is about 10% smaller than either the ΔH for Zn tracer diffusion or the ΔH for Cu tracer diffusion.[3] It seems probable that the relaxation is due to the movement of the more rapidly moving component; however, the exact mechanism is still unclear.

3-2. IMPURITY DIFFUSION IN PURE METALS

We turn now to the diffusion of substitutional impurities in pure metals. These experiments are performed by plating a very thin film of radioactive tracer on a pure metal, annealing, sectioning, and using the thin-film solution to determine D. This D is called the self-diffusion coefficient of the solute in the given solvent,[4] and the experiments are effectively done at infinite dilution.

In Sec. 2-3 it was shown that when a vacancy mechanism operates in a pure metal, the self-diffusion coefficient is determined by the fre-

[1] A discussion of the theory of this phenomenon is given by A. LeClaire and W. Lomer, *Acta Met.*, **2**: 731 (1954).

[2] M. deJong, *Acta Met.*, **10**: 334 (1962).

[3] J. Hino, C. Tomizuka, and C. Wert, *Acta Met.*, **5**: 41 (1957).

[4] The term "self-diffusion" has come to be applied to any diffusion coefficient measured with a tracer in the absence of a chemical concentration gradient. Strictly speaking, there is always a concentration gradient when A is plated on B or an A-B alloy, but with radioactive tracers the amount of solute added can be so small that the composition change can be ignored.

quency with which an atom will jump into a vacant neighboring site w and by the probability that a given neighboring site is vacant (p_v). In an infinitely dilute alloy, the problem is then to estimate whether, and by how much, w and p_v for a solute atom differ from w and p_v for a solvent atom. The treatments of this problem are all approximate and will be divided into two types. The first type ignores the difference in size between the solute and solvent and considers only the

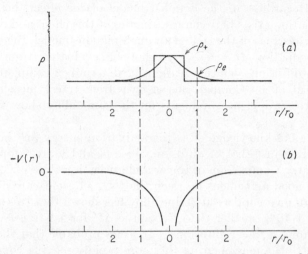

Fig. 3-6. (*a*) ρ_+ is the assumed positive charge density and ρ_e the resulting average electron density around a nucleus with a $+2$ charge in an array of nuclei with a $+1$ charge. r_0 is the nearest-neighbor distance. (*b*) shows the electrostatic potential resulting from the charge distribution of (*a*).

effects arising from valence differences, i.e., electronic interactions; the second ignores valence effects and considers only the difference in size between the two types of atoms.

Valence Effects.   Consider the case of a solvent of valence 1 and an impurity with the same size ion core but with a valence of 2, for example, cadmium in silver.   The model of the metal is a lattice of Ag^+ ions surrounded by a "gas" of free electrons.   If one of the silver atoms is removed and a cadmium atom put in its place, the cadmium atom will go to a Cd^{++} ion, thus contributing two electrons to the electron gas. The insertion of the cadmium ion gives a sharp change in the positive charge density ρ_+ (see Fig. 3-6a).   There is an electrostatic attraction between the impurity atom and the electron gas, which gives rise to a higher electron density ρ_e near the solute ion than far away from it. However, the electron density distribution is smooth around the impurity, since a sharp change in ρ_e would require high-energy electrons (short wavelengths) due to a quantum effect. The minimum value of the sum of the potential and kinetic energy of the electron gas is achieved if higher electron density does not completely "cancel out" or shield the extra charge of the Cd^{++} ion until a volume including nearest neighbors is considered. Figure 3-6a shows the variation of the positive charge density and the electron charge density around the cadmium ion. Loosely speaking, the ion's second electron is weakly bound to it and spends its time within one or two atomic diameters of the Cd^{++} ion. The ion is then said to be "screened." The electrostatic potential around the screened impurity can be approximated by the equation

$$V(r) = \frac{Ze}{r} \exp(-qr) \tag{3-8}$$

where Z is the number of excess electrons per ion (1 in the case of Cd) and r is the radial distance from the impurity. q is called the screening parameter and can be calculated from the free-electron theory of a metal.[1] Figure 3-6b shows $-V(r)$ around the screened impurity below the assumed positive charge density distributions. The potential $V(r)$ times the charge of an infinitesimal positive test charge is the energy of the test charge at r. The fact that $V(r)$ rises $[-V(r)$ falls] in the region $r < r_0$ results from the fact that a test charge would "see" and be repelled by the positive cadmium ion in this region. If it is assumed that the electrons stay uniformly distributed when a silver ion is removed to form a vacancy,[2] that is, ρ_e in Fig. 3-6 is not changed, the vacancy will have a charge of $-e$. Its energy on a site next to a Cd^{++} ion will then be reduced by

$$eV(r_0) = E(r_0) = \frac{Ze^2}{r_0} \exp(-qr_0) \tag{3-9}$$

Since the energy of a vacancy is lower next to a Cd^{++} ion, the concentration of vacancies on any given nearest-neighbor site will be increased to

$$p_v = N_v \exp\left[+\frac{E(r_0)}{kT}\right] \tag{3-10}$$

This makes ΔH_v on a site next to a solute less than ΔH_v on a site surrounded by solvent atoms, by $E(r_0)$. In addition to making $p_v > N_v$,

---

[1] The treatment described here was first given by D. Lazarus, *Phys. Rev.*, **93:** 973 (1954).

[2] It was stated in Sec. 2-7 that this would not be true. The excuse for assuming it here is that we are following Lazarus and the problem gets much more complicated if ρ_e is allowed to change.

the potential $V(r_0)$ makes ΔH_m for the solute less than ΔH_m for the solvent. This can be seen with the help of Fig. 3-6. We have agreed that the energy of a vacancy was decreased by $-eV(r_0)$ when the vacancy became a nearest neighbor of the solute. There is also a force $e(dV/dr)$ tending to draw the vacancy and the Cd^{++} ion together. If the Cd^{++} ion starts to move toward the vacancy, this force continues to pull the two together. This force decreases the activation energy by roughly $r_0 e(dV/dr)_{r=r_0}$. Rather arbitrarily taking the decrease in ΔH_m to be $-E(r_0)$, the total difference between the observed ΔH for cadmium and that for silver is given by the equation

$$\Delta H(\text{Cd}) = \Delta H(\text{Ag}) - 2E(r_0)$$

In this approximation, Z in Eq. (3-9) is the only quantity that changes with valence, and there are no disposable parameters. Thus for indium in silver, Z is increased by 1 and

$$\Delta H(\text{In}) = \Delta H(\text{Ag}) - 4E(r_0)$$

A plot of data for cadmium, indium, tin, and antimony in silver is shown in Fig. 3-7. The agreement for these elements is seen to be good. The same analysis should hold for zinc, gallium, germanium, and arsenic in copper, but Fig. 3-7 shows that the agreement with experiment is not as good. However, the theory is only a first-order approximation; the striking qualitative agreement indicates that conceptually the model is satisfactory for this type of impurity.

This same first-order theory should also apply for $Z < 0$. If it is ever possible to obtain a system in which $Z < 0$, it would be for solutes to the left of silver in the periodic table. It is indeed found that $\Delta H(\text{Ru}) > \Delta H(\text{Ag})$ or that ΔH for iron, nickel, or cobalt in copper is greater than $\Delta H(\text{Cu})$. However, the ambiguity of the valence for these elements in a solid solution is well known from phase-diagram work, so the use of the model for these solutes is questionable.[1]

Size Effects. We now wish to mention two other approaches which are applicable for estimating the change in ΔH_m if the effects due to the size of the solute are assumed to be much larger than those due to the relative valence of the solute and solvent.

The first approach is that of Swalin,[2] who assumes that the atoms are compressible spheres. For example, if a solute is larger than the solvent atom, ΔH_m will be increased by the additional strain required

---

[1] A bibliography and review of this theory is given by D. Lazarus, in F. Seitz and D. Turnbull (eds.), "Solid State Physics," vol. 10, p. 71, Academic Press, Inc., New York, 1960.

[2] R. Swalin, *Acta Met.*, **5**: 443 (1957).

at the saddle point. He considers the solute atom to be compressible and the lattice to be an elastic continuum. The change in ΔH_m is then a balance between the strain of the diffusing solute and the matrix. Since a large solute will elastically strain the lattice, it would seem that vacancies would be attracted to it, that is, ΔH_v is decreased. However, Swalin concludes that the change in ΔH_v is negligible. There are many approximations involved in such an approach, but the

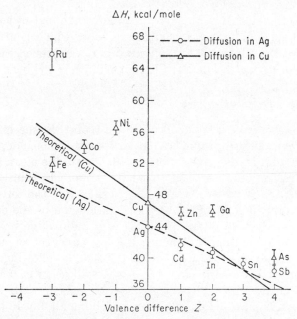

FIG. 3-7. Activation energy for diffusion (ΔH) in silver and copper versus Z for impurity. Straight lines are theoretical values. [*From D. Lazarus, in Seitz and Turnbull (eds.), "Solid State Physics," vol. 10, p. 71, Academic Press, Inc., New York*, 1960.]

resulting correlation between data and theory indicates that the concepts involved are useful ones.

The second approach is due to Overhauser.[1] In discussing the annealing out of low-temperature radiation damage (thought to be only excess vacancies and interstitials), he has considered the effect of interstitials on the activation energy for vacancy migration, that is, on ΔH_m. Since the interstitials dilate the lattice locally, the average interatomic distance between the matrix atoms will be increased. This expansion of the lattice will then reduce the work required to squeeze an atom through the saddle point, so he predicts that ΔH_m is decreased

[1] R. A. Overhauser, *Phys. Rev.*, **90**: 393 (1953).

by the addition of interstitials. The same type of argument predicts that a solute which expands the solvent lattice will decrease ΔH_m, while a solute which contracts the solvent lattice will increase ΔH_m. This effect will not change the solvent diffusion coefficient D_1 in extremely dilute solutions, as will the effects discussed above. However, Overhauser estimated that it would explain the factor of 2 increase in D_{Ag} upon the addition of 1 atom per cent lead, i.e., a decrease in ΔH of roughly 1.5 kcal.†

The three theories given above are no more than first-order approximations. Each points out a particular factor which contributes to making D for the solute and the solvent differ. However, each theory ignores the other factors, and a coherent picture is lacking.

3-3. CORRELATION EFFECTS

In all the cases we have studied up to this point the successive jump directions of the diffusing atom have been assumed to be independent of one another, that is, they are uncorrelated. Also, the mean frequency for all jumps has been assumed to be the same. In dilute substitutional alloys, this is not true for either the solute or the solvent.

This section will start with a detailed discussion of the correlation between the successive jump directions of a radioactive isotope of A in pure A. In this manner the existence and effect of correlation between successive jumps will be established. Following this, the effect of correlation on the equation for the diffusion coefficient of a substitutional impurity will be given and discussed. The correction for correlation in pure metals will be seen to be small, but in the case of impurity diffusion, it can be quite pronounced.

Pure Metals. The successive jump directions of an atom will be uncorrelated if after the nth jump all possible directions for the $(n + 1)$th jump are equally probable. For example, the jumps of a vacancy diffusing in a pure metal will be uncorrelated since after any jump all of the neighbors of the vacancy are identical. It follows that all possible jump directions have the same probability of occurring. This will not be the case for a tracer atom diffusing by a vacancy mechanism in a pure metal. After any jump of the tracer, all of its neighbors are not identical; one of them is a vacancy, and the most probable next jump direction for the tracer is right back to the site that is now vacant. This can be seen in the two-dimensional close-packed lattice in Fig. 3-8. If the tracer in the figure (at site 7) has just exchanged sites with the vacancy now at the site labeled 6, the

† R. Hoffman, D. Turnbull, and E. Hart, *Acta Met.*, **3**: 417 (1955).

tracer's most probable next jump is to return to 6.   Its next most probable jumps are to sites 1 and 5; i.e., the vacancy jumps from 6 to 5 (or 1) and then to 7.   The solute's least probable next jump is to 3, since this requires that the vacancy move from 6 around to 3 before it jumps to 7.   Since a return to 6 or a next jump to 1 or 5 would tend to cancel out the tracer's previous jump, the tracer will not travel as far in n jumps as a vacancy will.   Stated more exactly, the mean square displacement for the tracer after n jumps, $[\overline{R_n^2(t)}]$, will be less than that for a vacancy which took the same number of jumps, $\overline{R_n^2}(v) = n\alpha^2$. The degree of this canceling out is given by the correlation factor f which is defined by the equation

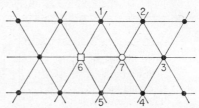

$$f = \lim_{n \to \infty} \frac{\overline{R_n^2}(t)}{\overline{R_n^2}(v)} \qquad (3\text{-}11)$$

Fig. 3-8. Portion of two-dimensional close-packed lattice showing a tracer (O) and a vacancy (□).

In the equations derived for D in Chap. 2, it was assumed that the atomic jumps were uncorrelated; this gave for the tracer diffusion coefficient $D_t = a_0^2 N_v w$.   It is now seen that instead

$$D_t = f a_0^2 N_v w$$

where $f < 1$ for diffusion by a vacancy mechanism.

In the following subsection the rigorous calculation of f will be discussed.   However, the calculation is complicated, and more insight can be gained initially from the rationalization of an approximate equation for f.   Returning to Fig. 3-8, the probability that the tracer atom will make its next jump back to 6 is just the same as the probability that the vacancy will jump from 6 to 7.   Since the vacancy jumps are random, the probability that the vacancy will jump from 6 back to 7 on its next jump is $\frac{1}{6}$, or in general $1/z$ where z is the number of nearest neighbors.   If the vacancy returns to site 7 on its next jump, this will have canceled out the first solute jump, and we can say that it has made two solute jumps ineffective.   Thus as a first approximation

$$f = 1 - \frac{2}{z} \qquad (3\text{-}12)$$

The insertion of $2/z$ in this equation is slightly arbitrary here, but it will be seen in the next subsection that this is the first, and largest. term in a series used to calculate f.

In Table 3-1 the values of f obtained from Eq. (3-12) are compared with the actual values of f calculated by Compaan and Haven.   A comparison of the second and third columns of the table shows that the values of f given by Eq. (3-12) are fairly close to the calculated values but are consistently too high. It is also seen that the values of f for the bcc and fcc lattices are fairly close to unity. Thus the insertion of f would not significantly change any of the calculations made, or conclusions drawn, in earlier chapters.

TABLE 3-1. Approximate and Actual Values of f for Vacancy Diffusion in Various Lattices

	z	$1 - 2/z$	f
Two-dimensional:			
Square................	4	0.5000	0.46705
Hexagonal.............	6	0.6667	0.56006
Three-dimensional:			
Diamond..............	4	$\frac{1}{2}$	$\frac{1}{2}$
Simple cubic...........	6	0.6667	0.65549
Body-centered cubic.....	8	0.7500	0.72149
Face-centered cubic......	12	0.8333	0.78145

SOURCE: K. Compaan and Y. Haven, *Trans. Faraday Soc.*, **52:** 786 (1956).

Calculation of f. To calculate f we shall first rearrange some of our random-walk equations. The equation derived for $\overline{R_n^2}$ in Sec. 2-3 was

$$\overline{R_n{}^2} = n\alpha^2 \left(1 + \frac{2}{n} \overline{\sum_{j=1}^{n-1} \sum_{i=1}^{n-j} \cos \theta_{i,i+j}} \right) \qquad (2\text{-}8)$$

If successive jump directions are independent of one another, all allowed values of θ are equally probable, and it was shown in Sec. 2-3 that $\overline{R_n{}^2} = n\alpha^2$. Substituting $\overline{R^2}(v) = n\alpha^2$ and Eq. (2-8) for $\overline{R^2}(t)$ in Eq. (3-11) gives

$$f = \lim_{n \to \infty} \left(1 + \frac{2}{n} \overline{\sum_{j=1}^{n-1} \sum_{i=1}^{n-j} \cos \theta_{i,i+j}} \right) \qquad (3\text{-}13)$$

The bar in this equation means to average the double sums for a large number of diffusing particles. The same quantity can be obtained by averaging the values of $\cos \theta$ for a large number of particles and summing these. That is, we can substitute

$$\overline{\sum_{j=1}^{n-1} \sum_{i=1}^{n-j} \cos \theta_{i,i+j}} \equiv \sum_{j=1}^{n-1} \sum_{i=1}^{n-j} \overline{\cos \theta_{i,i+j}} \qquad (3\text{-}14)$$

As a final aid in the reformulating of the equations, note that in any cubic or close-packed lattice all solute jumps are equivalent except for their orientation. That is, all tracer-vacancy pairs that have just completed an exchange are indistinguishable, aside from their orientation. Thus the value of $\overline{\cos \theta_{i,i+j}}$ is the same for each value of i. (If this is not apparent now, it should become clearer as the calculation of this term is explained.) Designating this new average as $\overline{\cos \theta_j}$, Eq. (3-13) can be written

$$f = \lim_{n \to \infty} \left[1 + \frac{2}{n} \sum_{j=1}^{n-1} (n-j) \, \overline{\cos \theta_j} \right] \tag{3-15}$$

In the limit as $n \to \infty$, $(n - j)/n = 1$ for the beginning terms of the series, so Eq. (3-15) can be written

$$f = 1 + 2 \, \overline{\cos \theta_1} + 2 \, \overline{\cos \theta_2} + \cdots \tag{3-16}$$

The problem is now reduced to calculating the mean value of the cosine of the angle between the ith and the $(i + 1)$th jump vectors $(\overline{\cos \theta_1})$, the mean value of $\cos \theta$ for the ith and the $(i + 2)$th jump vectors $(\overline{\cos \theta_2})$, etc.

As an example of the calculation of f, we shall outline the evaluation of f for the two-dimensional close-packed lattice shown in Fig. 3-8. To do this we must calculate $\overline{\cos \theta_1}$, the average value of the cosine of the angle between the last tracer jump vector (from 6 to 7) and the next tracer jump vector. For this it is necessary to calculate the probability of the tracer making its next jump to each of its six nearest neighbors. In general we define p_k as the probability that the tracer will make its next jump to its kth nearest neighbor. Similarly θ_k is the angle between the jump vector $6 \to 7$ and the jump vector $7 \to k$. The equation for $\overline{\cos \theta_1}$ is then

$$\overline{\cos \theta_1} = p_6 \cos \theta_6 + p_5 \cos \theta_5 + \cdots + p_1 \cos \theta_1$$

$$\overline{\cos \theta_1} = \sum_{k=1}^{z} p_k \cos \theta_k \tag{3-17}$$

The basic problem here (as well as in the case of impurities) is to calculate the various p_k. The probability that the vacancy will first return to site 7 from site k is just equal to p_k. The value of p_k is thus calculated by summing the various vacancy trajectories which will move the tracer to site k on the tracer's first jump. If we consider the case of the site labeled 6, that is, $k = 6$, p_6 can be calculated from the series

$$p_6 = n_{16}P_{16} + n_{26}P_{26} + \cdots = \sum_{i=1}^{m} n_{i6}P_{i6} \tag{3-18}$$

where P_{i6} is the probability that the vacancy will return for the first time to site 7 from site 6 on its ith jump. This then moves the *tracer* to 6 on its *first* jump. n_{i6} is the number of paths which will allow the vacancy to move the solute from 7 to 6 for the first time on its ith jump. Throughout this discussion, only one vacancy will be considered. Or equivalently, it is assumed that the density of vacancies is so low that no other vacancy will exchange with the tracer before the vacancy has randomized its position with respect to the initial solute-vacancy exchange.

Since the vacancy jumps are random, the probabilities of any particular vacancy path P_{i6} can be easily calculated. The probability that the vacancy will jump to any particular neighboring site on its first jump is $1/z$, or in this case $\frac{1}{6}$. The a priori probability that it will jump to one specified site and then to another specified site is $(\frac{1}{6})^2$. In general then $P_{i6} = (\frac{1}{6})^i$.

The first few values of n_{i6} can be obtained by inspection. There is just one path for the vacancy which will bring it back to 7 on its first jump, so $n_{16} = 1$. There are no paths which will return the vacancy to 7 from 6 on its second jump, so $n_{26} = 0$. The value of n_{36} is five since a first jump by the vacancy to any of its nearest neighbors, aside from the tracer, and a return to 6 on the next jump will allow the vacancy to go to 7 from 6 on its third jump. A similar examination shows that $n_{46} = 8$. The calculation of higher values of n_{i6} becomes tedious and difficult, yet n_{i6} increases with i, so n_{i6} must be obtained for relatively large values of i to obtain accurate values of $\overline{\cos \theta_i}$. Our aim here is only to outline the calculation, so we shall only work with terms in $i \leq 4$.[†] Carried out to the fourth jump of the vacancy, that is, $i = 4$, in Eq. (3-18), we obtain

$$p_6 = \tfrac{1}{6} + 0 + 5(\tfrac{1}{6})^3 + 8(\tfrac{1}{6})^4 = 0.1960$$

By symmetry p_5 must equal p_1. To calculate these we note that $n_{15} = n_{11} = 0$, since the vacancy cannot return to 7 from 1 on its first jump. The equation for p_5 or p_1 carried out to the fourth vacancy jump is

$$p_5 = p_1 = 0 + (\tfrac{1}{6})^2 + (\tfrac{1}{6})^3 + 11(\tfrac{1}{6})^4 = 0.0409$$

Similarly

$$p_4 = p_2 = 0 + 0 + 1(\tfrac{1}{6})^3 + 2(\tfrac{1}{6})^4 = 0.0062$$
$$p_3 = 2(\tfrac{1}{6})^4 = 0.0015.$$

This then shows that the probability of the tracer in Fig. 3-8 making its next jump back to site 6 (p_6) is over 100 times the probability

[†] The details of calculating $\overline{\cos \theta_j}$ for many cases are given by K. Compaan and Y. Haven, *Trans. Faraday Soc.*, **52**: 786 (1956).

that it will make its next jump in the same direction as the first, i.e., to site 3. Substituting these values of p_k into Eq. (3-17)

$$\overline{\cos \theta_1} = (-1)0.196 + (-\tfrac{1}{2})(2)0.0378 + (\tfrac{1}{2})(2)0.0061$$
$$+ (1)0.0015 = -0.2262$$

The term in $\overline{\cos \theta_2}$ takes account of the correlation between the ith jump and the $(i + 2)$th jump of the solute. Compaan and Haven show that $\overline{\cos \theta_j} = (\overline{\cos \theta_1})^j$ for a vacancy mechanism so that Eq. (3-16) can be written

$$f = 1 + 2 \overline{\cos \theta_1} + 2(\overline{\cos \theta_1})^2 + 2(\overline{\cos \theta_1})^3 + \cdots \qquad (3\text{-}19)$$

But this infinite series can be replaced by

$$f = \frac{1 + \overline{\cos \theta_1}}{1 - \overline{\cos \theta_1}} \qquad (3\text{-}20)$$

as can be shown by dividing $1 + \overline{\cos \theta}$ by $1 - \overline{\cos \theta}$.

Using the first three figures of the value of $\overline{\cos \theta_1}$ calculated here gives $f = 0.631$. This is to be compared with the value $f = 0.560$ in Table 3-1. The disagreement stems entirely from the omission of vacancy paths involving $i > 4$. The fraction of the possible vacancy trajectories which has been omitted can be seen by noting that the sum of all return probabilities $\left(\sum\limits_{k=1}^{6} p_k \right)$ is only 0.286 instead of unity. Since the probability is unity that the tracer will make a next jump by exchanging with the vacancy initially at site 6, this means that Σp_k would increase over 70% if terms in higher values of i were included in the calculation of each p_k.[†] Although 70% of the returns is a relatively large omission, it is seen that the value of f calculated here would take care of 84% of the total correlation. The main conclusion to be drawn is that Σp_k and f converge very slowly to their final values as i is increased. Hastening this convergence is the main problem in calculating f.

[†] That Σp_k will in fact equal 1 after a finite number of vacancy jumps can be shown by appealing to the continuum treatment of Chap. 1. If an instantaneous point source is placed in a two-dimensional medium at $t = 0$, the solution to the diffusion equation is $c(r,t) = (\text{const}/t) \exp (-r^2/4Dt)$. Thus the probability of finding a vacancy at its initial site in some time increment dt after a time t is $p(t)\, dt = (\text{const}/t)\, dt$. Now the probability that the vacancy will return to its initial site at some time between $t = \epsilon$ and $t = t$ is $\int_\epsilon^t p(t)\, dt = \ln t \Big]_\epsilon^t$. But this equals unity after some finite time. Therefore the vacancy will return to the tracer after some finite time.

Let us next consider the justification of the approximate equation

$$f = 1 - \frac{2}{z} \tag{3-12}$$

given in the preceding subsection. If only the probability of the vacancy exchanging with the tracer on its first jump is considered [$i = 1$ in Eq. (3-18)], then $p_6 = -1/z$, and all other $p_k = 0$. This gives $\overline{\cos \theta_1} = -1/z$. If this is combined with the first two terms in Eq. (3-19), Eq. (3-12) results.

The calculation of f for a three-dimensional lattice is completely analogous to the calculation outlined here. The added dimension makes the problem more difficult but no different in principle. The values of f for various cubic lattices are given in Table 3-1.

In closing, it should be emphasized that the existence of correlation between successive jumps does not mean that out of many jumps for the same particle one jump vector occurs less frequently than another. If the many jump vectors of a tracer were tabulated, it would be seen that all jump vectors occurred with the same frequency, just as in the case of a diffusing vacancy. The fact that $\overline{R_n^2}(t) < n\alpha^2$ stems exclusively from the order in which the jump vectors occur.

Very Dilute Alloys. One might ask "So what?" to the detailed calculation of the factor f given above, since the basic equation is unchanged and we cannot calculate Γ for the tracer to within a factor much larger than $1/f$. However, the effect is more easily treated for pure metals, and in the case of impurity diffusion an understanding of correlation effects is fundamental to even a qualitative discussion of just what atomic processes determine the diffusion coefficient for the impurity.

In alloys, vacancies interact with impurities, so not only are the jumps of the impurities correlated, but the successive jumps of the vacancies are too. As an extreme example, consider the diffusion of an impurity which is "bound" or strongly attracted to a vacancy. To simplify the geometry, we again consider a two-dimensional close-packed lattice. Assume that the vacancy-impurity exchange rate w_2 is much greater than the rate at which the vacancy exchanges with a solvent atom which is also a nearest neighbor of the impurity w_1. Under these conditions the successive jump directions of the vacancy are no longer random but will be almost completely correlated. That is, if the vacancy exchanges with the impurity, the probability that it will reverse that jump on its next exchange is almost unity. For this reason the impurity will translate through the lattice only as fast as the vacancy exchanges with the solvent atoms. Under these conditions the impurity diffusion coefficient will be given not by the equation

$$D_2 \simeq a^2 w_2$$

but by the equation[1]

$$D_2 \simeq a^2 w_1 = f a^2 w_2 \ll a^2 w_2 \tag{3-21}$$

Here $f \ll 1$, and it is obvious that correlation effects play a dominant role in determining D_2.

To be more quantitative, an approximate equation for f can easily be derived with the equations of the preceding subsection. We first assume that the vacancy-impurity pair will not dissociate and that for its last jump the impurity exchanged with the vacancy shown in Fig. 3-9. The calculation of $\overline{\cos \theta_1}$ is parallel to the case in which the vacancy jumps were random, except that in place of $1/z$ the probability that the vacancy and the impurity will exchange again on the next vacancy jump is $w_2/(w_2 + 2w_1)$.† If we consider only this immediate vacancy-impurity exchange in calculating $\overline{\cos \theta_1}$, we obtain $p_6 = w_2/(w_2 + 2w_1)$. The approximate equations are thus

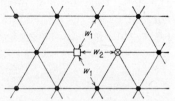

Fig. 3-9. Two-dimensional close-packed lattice showing vacancy ($\square$), impurity atom ($\otimes$), and solvent atoms ($\bullet$). This differs from the case shown in Fig. 3-8 in that the vacancy-impurity exchange rate w_2 and the vacancy-solvent exchange rate w_1 are different.

$$\overline{\cos \theta_1} = - \frac{w_2}{w_2 + 2w_1}$$

$$f = \frac{1 + \overline{\cos \theta_1}}{1 - \overline{\cos \theta_1}} = \frac{w_1}{w_2 + w_1} \tag{3-22}$$

Ignoring geometric constants, the equation for D_2, the diffusion coefficient of the solute atom, is

$$D_2 \simeq \frac{a_0^2 w_2 w_1}{w_1 + w_2} \tag{3-23}$$

[1] Here and in what follows, the subscript 1 will refer to the solvent atoms while the subscript 2 will refer to the impurity.

† To show this, note that the probability that the vacancy and the impurity will exchange in the time element δt is proportional to $w_2 \delta t$ if $\delta t \ll 1/w_2$. The probability that the vacancy will make any one of the three possible jumps in δt is proportional to $(2w_1 + w_2)\delta t$. The probability of the next vacancy exchange being with the impurity is the ratio of these two or $w_2/(w_2 + 2w_1)$.

Although this equation is only a first approximation, it has all of the characteristics of the exact solution for two (or three) dimensions. Inspection shows that D_2 is determined primarily by the slower of the two jump frequencies. If the impurity-vacancy exchange is much faster ($w_2 \gg w_1$), then $D_2 \simeq a_0^2 w_1$. If the reverse is true ($w_2 \ll w_1$), then $D_2 \simeq a_0^2 w_2$. Finally, if $w_2 = w_1$ then $D_2 \simeq a_0^2 w_2/2$.

The reason that D_2 is less than $a_0^2 w_2$ in this final case can be understood by studying Fig. 3-9. Before the solute can move directly to the left, the vacancy must make three successive exchanges with solvent atoms (w_1-type jumps) to get to the left of the solute. If, after this occurs, the vacancy and the impurity exchange, three w_1-type jumps must occur again before a w_2-type jump will move the solute to the left. Thus both w_1 and w_2 jumps must occur for the impurity-vacancy pair to diffuse, and if $w_1 = w_2$, the pair moves with a frequency roughly equal to $w_2/2$.

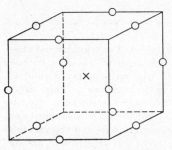

Fig. 3-10. Impurity atom ($\times$) and nearest neighbors ($\bigcirc$) in an fcc lattice.

The behavior in a three-dimensional lattice is just the same. If an impurity-vacancy pair is formed in an fcc lattice, the only difference from the above example is that there are four solvent atoms which the vacancy can move to without dissociating the impurity-vacancy pair, instead of two (see Fig. 3-10). The probability of an impurity-vacancy exchange on the next vacancy jump is thus $w_2/(4w_1 + w_2)$. The exact equation for f is found to be

$$f = \frac{w_1}{w_1 + w_2} \tag{3-24}$$

To obtain an exact equation for D_2 note that the basic equation for D_2 is

$$D_2 = a_0^2 f w_2 p_v \tag{3-25}$$

Since it has been assumed that the vacancy-impurity pairs do not dissociate, the impurity always has a vacancy on one of its twelve neighboring sites, that is, $p_v = \frac{1}{12}$. Therefore

$$D_2 = \frac{a_0^2}{12} \frac{w_1 w_2}{w_1 + w_2} \tag{3-26}$$

This equation varies with the ratio w_1/w_2 just as Eq. (3-23) did.

For the case in which the vacancy-impurity pair is less tightly bound, the rate of dissociation of the pairs must be included. If we assume that the region affected by an impurity includes only its nearest neighbors, k_1 can be taken as the frequency with which an associated vacancy exchanges with one of the seven solvent atoms which is not a nearest neighbor of the impurity atom. The probability of an impurity-vacancy pair dissociating on its next jump is $7k_1/(w_2 + 4w_1 + 7k_1)$. If we calculate an approximate value of f as before by assuming $\overline{\cos \theta_1} = -w_2/(w_2 + 4w_1 + 7k_1)$, we obtain

$$f = \frac{1 + \overline{\cos \theta_1}}{1 - \overline{\cos \theta_1}} = \frac{2w_1 + 7k_1/2}{w_2 + 2w_1 + 7k_1/2} \tag{3-27}$$

A more complete calculation, but one which ignores the return of the dissociated vacancies, gives[1]

$$f = \frac{w_1 + 7k_1/2}{w_1 + w_2 + 7k_1/2} \tag{3-28}$$

Substituting this in Eq. (3-25) gives

$$D_2 = \frac{a_0^2 w_2(w_1 + 7k_1/2)}{w_1 + w_2 + 7k_1/2} \, p_v \tag{3-29}$$

The type of complete or partial association of impurities and vacancies discussed above is important in the study of impurity diffusion in ionic materials. For example, if a divalent impurity, such as Mg^{++}, is dissolved in $NaCl$, the magnesium ion will take the place of a sodium ion on a lattice site. However, to maintain charge neutrality, another sodium ion must be removed from the lattice; that is, a vacant sodium ion site must be created. There will be a strong electrostatic (coulombic) attraction between the Mg^{++} ion and the sodium vacancy, so the fraction of magnesium ions paired with vacancies will be high. Thus it is not hard to see why this problem was first treated for the case of impurity diffusion in ionic materials.[2]

From the above, it is also apparent that the simple treatment of the effect of the impurity-vacancy interaction on D_2 discussed in Sec. 3-2 is at best an approximation of unknown accuracy. The discussion given there allowed the estimation of p_v and w_2; however, this still leaves w_1 and k_1 to be calculated. Since all these frequencies cannot now be calculated, the best we can do is to consider what determines D_2 in several limiting cases.

[1] For a relatively simple means of calculating f and a list of older references, see H. B. Huntington and P. B. Ghate, *Phys. Rev. Letters*, **8**: 421 (1962).

[2] A. Lidiard, *Phil. Mag.*, **46**: 1218 (1955).

1. If the rate of dissociation of vacancy-impurity pairs $(7k_1/2)$ is much less than the w_1, Eq. (3-29) simplifies to

$$D_2 \simeq a_0^2 \frac{w_2 w_1}{w_1 + w_2} p_v \qquad (3\text{-}30)$$

which is essentially Eq. (3-26).

2. If the impurity-vacancy exchange is by far the most rapid, then $w_2 \gg (w_1 + 7k_1/2)$ and

$$D_2 \simeq a_0^2 \left(w_1 + \frac{7k_1}{2} \right) p_v \qquad (3\text{-}31)$$

3. If the impurity atom jumps relatively slowly, $w_2 \ll (w_1 + 7k_1/2)$ and

$$D_2 \simeq a_0^2 w_2 p_v \qquad (3\text{-}32)$$

4. Finally, if the "impurity" is a solvent tracer, then

$$w_1 = w_2 = k_1 = w_0$$

and $p = N_v$. Thus

$$D_2 = \frac{9}{11} a_0^2 N_v w_0 \qquad (3\text{-}33)$$

Note that in this last equation $f = \frac{9}{11} = 0.818$, even though correlation effects were only approximated in the derivation.

Table 3-2 summarizes much of the experimental data on D_2 at infinite dilution. It is seen that D_1 can be greater than or less than D_2, but rarely differs by more than a factor of 10. Looking back over Eqs. (3-25) to (3-32), considerable insight can be gained into the interplay of effects which determine D_2. However, upon attempting to determine experimentally the relative values of w_1, w_2, and k_1 from these data, it soon becomes apparent that one has two unknown ratios and only one experimentally measurable quantity. Thus we are left in a quandary. To obtain additional measurable quantities, one can resort to measurements of D_1, but this introduces w_0, the jump frequency of the solvent atom in the pure solvent. If D_1 and D_2 are studied as a function of N_2 for $N_2 \ll 1$, the quantities dD_1/dN_2 and dD_2/dN_2 are obtained, and no new jump frequencies are introduced. However, the detailed treatment of D_1 in a manner similar to that given for D_2 is even more complex. Also the analysis shows that there are still not enough measurable quantities unless further approxima-

TABLE 3-2. Impurity Diffusion Coefficients in Pure Copper and Silver

Solvent	Impurity	Ref.	$D_0 \left(\dfrac{cm^2}{sec}\right)$	$\Delta H \left(\dfrac{kcal}{mole}\right)$	D_2/D_1 (1000°K)
Silver.........	Pb	1	0.22	38.1	11.0
	Sb	2	0.17	38.3	7.8
	Sn	3	0.25	39.3	6.9
	In	3	0.41	40.6	5.7
	Zn	4	0.54	41.7	4.4
	Hg	4	0.08	38.1	4.0
	Cd	3	0.44	41.7	3.7
	Tl	5	0.15	39.9	3.1
	Cu	4	1.2	46.1	1.1
	Ag	7	0.40	44.1	1.00
	Au	4	0.26	45.5	0.25
	Ru	6	180	65.8	0.01
Copper........	As	8	0.12	42.0	7.6
	Hg	8	0.35	44.0	5.5
	Ga	8	0.55	45.9	5.0
	Ag	8	0.63	46.5	4.2
	Zn	9	0.34	45.6	3.6
	Cu	11	0.20	47.1	1.00
	Au	8	0.69	49.7	0.95
	Fe	10	1.4	51.8	0.67
	Co	10	1.93	54.1	0.29
	Ni	10	2.7	56.5	0.12

[1] R. E. Hoffman, D. Turnbull, and E. W. Hart, *Acta Met.*, **3**: 417 (1955).

[2] E. Sonder, L. Slifkin, and C. Tomizuka, *Phys. Rev.*, **93**: 970 (1954).

[3] C. T. Tomizuka and L. M. Slifkin, *Phys. Rev.*, **96**: 610 (1954).

[4] A. Sawatsky and F. Jaumot, Jr., *Trans. AIME*, **209**: 1207 (1957).

[5] R. E. Hoffman, *Acta Met.*, **6**: 95 (1958).

[6] C. B. Pierce and D. Lazarus, *Phys. Rev.*, **114**: 686 (1959).

[7] C. T. Tomizuka and D. Lazarus, *J. Appl. Phys.*, **25**: 1443 (1954).

[8] C. Tomizuka, quoted by D. Lazarus, "Solid State Physics," vol. 10, p. 117, Academic Press, Inc., New York, 1960.

[9] J. Hino, C. Tomizuka, and C. Wert, *Acta Met.*, **5**: 41 (1957).

[10] C. A. Mackliet, *Phys. Rev.*, **109**: 1964 (1958).

[11] A. Kuper, H. Letaw, L. Slifkin, E. Sonder, and C. Tomizuka, *Phys. Rev.*, **96**: 1224 (1954); errata *ibid.* **98**: 1870 (1955).

tions are made. Some of these results will be summarized in the next section.

3-4. DIFFUSION IN DILUTE BINARY ALLOYS

In the preceding section we discussed the reasons for D_2 differing from D_1. In this section we shall discuss the variation of the solvent and solute diffusion coefficients with the solute concentration N_2. This problem has been treated by several authors. In order of increasing detail, these treatments are given below.[1]

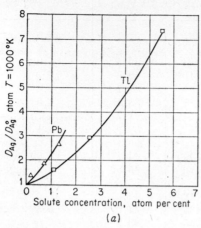

(a)

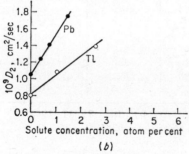

(b)

FIG. 3-11. (a) Variation of D_{Ag} with solute concentration for lead and thallium solute additions. [*From R. Hoffman, Acta Met.*, **6**: 95 (1958).] (b) Variation of solute self-diffusion coefficient in silver (D_2) for lead and thallium at 1000°K. [*Data of R. Hoffman, Acta Met.*, **6**: 95 (1958).]

In a very dilute alloy most of the solvent atoms are not near solute atoms, and D_1 is given by the equation

$$D_1 = a_0{}^2 w_0 N_v \qquad (3\text{-}34)$$

where w_0 is the jump frequency in the pure solvent. However, as N_2 increases, this will change. Some general characteristics of the variation of D_1 with N_2 can be seen from a qualitative consideration of Eqs. (3-33) and (3-29). If $p_v = N_v$ and $w_0 \simeq w_1 \simeq k_1$, then no matter how large w_2 is, D_2 will be about equal to D_1. If D_2 is greater than D_1 in an infinitely dilute solution, it is apparent that either the solute attracts vacancies so that $p_v > N_v$ near a solute or the *solvent* jump frequencies w_1 and k_1 are increased in the neighborhood of the solute. But if $p_v > N_v$ or $w_1 > w_0$, then D_1 is greater in the neighborhood of a solute atom than away from a solute atom. As N_2 increases, so does the number of solvent atoms affected. It follows then that if $D_2 > D_1$, the addition of solute will increase D_1, that is, $dD_1/dN_2 > 0$. Conversely, if $D_2 < D_1$, then dD_1/dN_2 will be negative.

For dilute solutions an estimate of the magnitude of this increase can be obtained as follows. We assume that the diffusion coefficient of the solvent is made up of two parts—that due to solvent atoms

[1] R. Hoffman, E. Hart, and D. Turnbull, *Acta Met.*, **3**: 417 (1955) and **5**: 74 (1957). D. Lazarus, in F. Seitz and D. Turnbull (eds.), "Solid State Physics," vol. 10, p. 71, Academic Press, Inc., New York, 1960. H. Reiss, *Phys. Rev.*, **113**: 1445 (1959).

which are not influenced by a solute and that due to solvent atoms which are influenced by neighboring solute atoms. We shall assume that this "realm of influence" of the solute atom contains only its nearest neighbors. Since each solute has 12 nearest neighbors, the fraction of solvent atoms which are not affected by solute atoms is $1 - 12N_2$. The unaffected atoms will diffuse at a rate equal to that of D_1 in a pure metal [designate $D_1(0)$]. The fraction of solvent which is influenced by the solute will be roughly $12N_2$. It seems reasonable to say that D_1 in these regions is about equal to D_2, and it can be shown that this is true. From this simplified argument, we arrive at approximate equations which are valid for very dilute solutions, that is, $N_2 \leq 0.01$,

$$D_1(N_2) = (1 - 12N_2)D_1(0) + 12\beta N_2 D_2(N_2) \qquad (3\text{-}35)$$
$$D_2(N_2) = D_2(0) \qquad (3\text{-}36)$$

where β is a constant ≤ 1. As might be expected, the actual derivation of the constant β is complicated, and its value varies from system to system. Nevertheless, Eq. (3-35) does fit the results quite satisfactorily for the case of $D_1 > D_2$ and for $D_1 < D_2$. This can be seen for two examples of $D_1 < D_2$ in Fig. 3-11.[1]

The lines drawn in the top half of Fig. 3-11 are those obtained from Eqs. (3-35) and (3-37) with $\beta = 0.6$ for lead and $\beta = 0.4$ for thallium. The data given are not in the very dilute region, and it is seen that D_2 definitely increases with N_2. Thus it is found empirically that Eq. (3-36) must be replaced by an equation of the form

$$D_2(N_2) = D_2(0)(1 + \alpha N_2) \qquad (3\text{-}37)$$

in the range $N_2 \geq 0.01$. This equation can be rationalized without recourse to a detailed derivation. We have assumed that the presence of a solute changes the probability that a given site will be vacant (p_v) and changes the jump frequencies for its nearest-neighbor sites. Thus, if the nearest-neighbor shells of two solute atoms overlap, D_2 will be affected. The probability that there will be a second solute atom within the second-nearest-neighbor shell of the first is proportional to N_2. Also, when $N_2 > 0.01$, the probability of two such shells overlapping is appreciable. Thus the variation of $D_2(N_2)$ shown in Fig. 3-11 is understandable. This problem has been discussed briefly by Reiss. The effect also becomes more apparent if the model of Overhauser is kept in mind (Sec. 3-2).

[1] Data for the case of $D_1 > D_2$ are quite rare in silver. Palladium in silver is the only case that has been studied in detail [Nachtrieb, Petit, and Wehrenberg, *J. Chem. Phys.*, **26**: 106 (1957)]. For a discussion of the interpretation, see H. Reiss, *Phys. Rev.*, **113**: 1445 (1959).

PROBLEMS

3-1. Given an internal friction peak in a bcc metal which is due to interstitial diffusion, at 0.7 cycle/sec the maximum in the internal friction plot occurs at 100°C. If the frequency is changed to 1.8 cycles/sec the peak is shifted to 120°C. Calculate ΔH for the process. If $a_0 = 3.2$ A, calculate D at 100°C and D_0.

3-2. Draw a fcc unit cell, and explain why individual interstitial atoms do not give rise to an internal friction peak. Could interstitial pairs give rise to internal friction?

3-3. In tantalum, the interstitial oxygen atoms attract each other with an interaction energy of about -0.1 ev. Also, the interstitials in a pair have a lower ΔH for jumping than isolated interstitials. To give the maximum separation of the peaks (on a plot of tan δ versus $\omega\tau$) would you work with a torsion pendulum or the elastic aftereffect? Why?

3-4. Would you expect any correlation between the jump vectors of a tracer in a pure metal if diffusion occurs by
 (*a*) A ring mechanism?
 (*b*) An interstitial mechanism?
 (*c*) An interstitialcy mechanism?

3-5. Calculate $\overline{\cos \theta_1}$ and f for a tracer diffusing in a two-dimensional square lattice. (Consider only vacancy trajectories which move the solute again in three or less vacancy jumps.)

3-6. In a two-dimensional square lattice with a solvent tracer diffusing by an interstitialcy mechanism
 (*a*) Calculate $\overline{\cos \theta_1}$ if the tracer starts as the interstitial atom as shown in Fig. 3-12.

FIG. 3-12   FIG. 3-13

 (*b*) Calculate $\overline{\cos \theta_1}$ if the tracer starts on a normal site (see Fig. 3-13). (Assume that an atom pushed from a normal site has an equal probability of going to each of the three available interstitial sites.)
 (*c*) Show that $f = 1 + 2 \overline{\cos \theta}$ for an interstitialcy mechanism where $\overline{\cos \theta}$ is the value calculated in part (*b*).

3-7. Calculate and plot f for an impurity-vacancy pair in an fcc lattice for $w_1/w_2 = 10, 5, 1, \frac{1}{5},$ and $\frac{1}{10}$.

3-8. Calculate $\cos \theta_1$ for the impurity-vacancy pair of Fig. 3-9. (Consider only vacancy trajectories which move the solute again in three or less vacancy jumps.)

3-9. Over a decade before correlation effects were treated in the manner described in the text, R. P. Johnson [*Phys. Rev.*, **56**: 815 (1939)] derived equations for the diffusion coefficient of an impurity atom when it does and when it does not attract a vacancy. Give his equations and assumptions in our terminology, and compare his results and ours. Is his treatment for these two cases valid?

chapter 4 DIFFUSION IN A
CONCENTRATION GRADIENT

In the preceding chapter our discussion of diffusion in substitutional alloys was limited to self-diffusion experiments. In such experiments the specimen is, or is assumed to be, chemically homogeneous. Such studies showed that the self-diffusion coefficients are in general different for the two elements in a substitutional alloy. Yet, if two semi-infinite bars of differing proportions of components 1 and 2 are joined and diffused, the Boltzmann-Matano solution gives only one diffusion coefficient $D(c)$ which completely describes the resulting homogenization. Thus the problem is to relate this single diffusion coefficient at a given composition to the self-diffusion coefficients at the same composition. To do this there are two new effects which must be understood. The first of these concerns the kind of matter flow which is to be classified as diffusion. In a binary diffusion couple with a large concentration gradient we shall see that diffusion gives rise to the shifting or flow of one part of the diffusion couple relative to another. Since the diffusion coefficient resulting from the Boltzmann-Matano solution is given by the equation[1]

$$\tilde{D} = -\frac{J}{dc/dx} \qquad (4\text{-}1)$$

[1] In this chapter it is necessary to work with a large number of diffusion coefficients, all of which apply to the same system but which are different. There are no generally accepted names for these so we shall define each of them by an equation. The D obtained from the Boltzmann-Matano solution is often called the chemical diffusion coefficient and will be designated $\tilde{D}$.

115

any shifting or flow of lattice planes relative to the ends of the diffusion couple is recorded as a flux and affects $\tilde{D}$ even though it does not correspond to any jumping of atoms from one site to another.

The second effect stems from differences in the actual diffusion processes going on in self-diffusion studies and in a couple with a large concentration gradient. It will be seen that these differences arise from both the nonideality of the alloy and the difference in the correlation effects in the two experiments. The development will be based primarily on thermodynamic or phenomenological reasoning as opposed to the mechanistic, or atomistic, models of Chaps. 2 and 3.

4-1. THE KIRKENDALL EFFECT

Intuitively it would seem that $\tilde{D}$ is some kind of mean value of the diffusion coefficients of components 1 and 2. But, if components 1

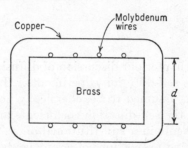

FIG. 4-1. Schematic diagram showing a cross section of the diffusion couple used by Smigelskas and Kirkendall.

and 2 diffuse at different rates in a binary alloy, it is necessary to obtain some parameter other than $\tilde{D}$ which indicates the magnitude of this difference. The first experiment which allowed the determination of this difference for alloys was discovered by Kirkendall.[1] In the experiment used, a rectangular bar of 70-30 brass was wound with fine molybdenum wire (molybdenum is insoluble in copper and brass) and then plated with about 0.1 in. of pure copper. This couple was then given a series of successive anneals. After each anneal, a piece was cut from the bar, sectioned, and the distance between the Mo wires (d) was measured (see Fig. 4-1). It was found that d decreased monotonically with time. In the Cu-Zn system there is a volume change on adding copper to brass, but even after this effect was subtracted out, a definite marker shift remained. This shift required that the flux of zinc atoms outward past the markers be appreciably

[1] A. Smigelskas and E. Kirkendall, *Trans. AIME*, **171**: 130 (1947).

greater than the flux of copper atoms inward across the same plane. Kirkendall had attempted to show this effect in two earlier papers, and in this case the results were sufficiently unequivocal to move his peers. In 1947 this was a new concept,[1] and its generality was not apparent. However, later work on a variety of markers in many different alloy systems confirmed these results, and the effect has proved to be quite general.[2]

4-2. DARKEN'S ANALYSIS

In 1948 Darken published an analysis of diffusion in alloys which was inspired by the experiments of Smigelskas and Kirkendall.[3] In it he establishes answers to the question of how the self-diffusion coefficients are related to $\tilde{D}$ and the nonideality of the alloy. The analysis has withstood the test of time, and most of the chapter will be devoted to a discussion of it. Darken's original paper is an excellent example of a phenomenological or thermodynamical analysis. The basic characteristic of this approach is that no mechanism or model is assumed. The student should develop a feeling for the relative advantages and disadvantages of this type of analysis as compared with a mechanistic analysis. With this in mind, the reader is urged to read Darken's original paper and the very interesting discussion that follows it.

Darken starts his paper by discussing the question, "What is diffusion?" The marker movement experiments show that the region around the marker translates relative to the region in which no diffusion occurs. The uniform translation or flow of an entire region across a reference plane gives a flux through the plane, but this is not what we would choose to call a diffusion flux across the plane. Thus we must define a new coordinate system which will allow us to measure only what we choose to call atomic diffusion. The concepts of flow may be clearer if the problem of the diffusion of ink in a moving stream

---

[1] While the vacancy mechanism was accepted for diffusion in pure fcc metals in 1947, it was thought that both constituents diffused at the same rate in alloys. If they were not the same, it was not at all clear what the mechanism was which would allow the two components to diffuse at different rates. The reader who is interested in the change which this made in concepts of alloy diffusion will find Smigelskas and Kirkendall's paper and its discussion interesting reading.

[2] The work of L. C. Correa da Silva and R. Mehl, *Trans. AIME,* **191:** 155 (1951), is one of the careful, early confirmations. For a more complete list of such references, see D. Lazarus, in Seitz and Turnbull (eds.), "Solid State Physics," vol. 10, p. 71, Academic Press, New York, 1960.

[3] L. Darken, *Trans. AIME,* **174:** 184 (1948).

of water is considered. If water containing a concentration c of ink flows past a point on the bank at a velocity v, the flux of ink past that point will be vc plus any flux due to a concentration gradient in the water. If we wish to study only this latter contribution to the flux, it is necessary to determine v. To do this we place chips on the water and take their velocity to be that of the water. In order to be sure that the chips do indeed reflect the velocity of the water, we can place chips of several different sizes and materials on the water. If they all move at the same speed, we can safely assume that the velocity of the water corresponds to that of the chips. If we now want to study the diffusion of ink in this stream, we can do so in the neighborhood of a chip by subtracting the quantity vc from the flux relative to the bank or by measuring the flux of ink in a coordinate system fixed relative to the chips.

In metals it is found that markers of different material, e.g., tungsten wires, molybdenum wires, or oxide inclusions, move at the same rate, so we can take diffusion to be the net flux relative to a nearby marker. It is quite probable that the markers are fixed relative to the lattice planes, but Darken chose not to define his flux relative to lattice planes since there is no operation which allows one to pick out a particular plane of atoms before and after diffusion. Thus, following Bridgman's recommendations, he used a coordinate system whose origin was operationally defined,[1] i.e., inert markers.

To set up the required equations, consider a diffusion couple made by joining two semi-infinite homogeneous rods containing different proportions of components 1 and 2. Along the direction of the diffusion flux will be placed many small, inert particles which act as markers. We can define the u_1 axis as being parallel to the diffusion direction and having its origin fixed on the number 1 marker. From the discussion of ink diffusion in water, any flux relative to the u_1 axis in the vicinity of $u_1 = 0$ is said to be due to diffusion. For a unit area this flux will be

$$-D_1 \frac{\partial C_1}{\partial u_1}$$

[1] P. Bridgman has pointed out that the revolution which rocked the science of physics with the advent of relativity and quantum mechanics might have been avoided if physicists had worked only with quantities which they could define by prescribed operations. "Hence arises the demand that the concepts or terms used in the description of experience be framed in terms of operations which can be unequivocally performed." "Reflections of a Physicist," p. 8, Philosophical Library, Inc., New York, 1955. Those interested in this philosophy are referred to P. Bridgman, "The Logic of Modern Physics," The Macmillan Company, New York, 1932.

where C_1 is in moles per unit volume. Since we are treating the case in which markers move relative to one another, the flux of material across a plane fixed relative to the u_1 axis will not be given by this expression in regions far from the plane $u_1 = 0$. In general, we can discuss diffusion in any given region by defining an appropriate u_n axis which is fixed relative to the marker in this region (designated the nth marker). However, many coordinate systems are awkward to work with. To avoid this, consider a new coordinate axis x† which is parallel to the various u_i axes, but may be moving relative to them. Since both of the derivatives $(\partial C_1/\partial u_n)_t$ and $(\partial C_1/\partial x)_t$ are taken at a given instant, the relative movement of x and u_n does not enter the problem, and $(\partial C_1/\partial u_n)_t = (\partial C_1/\partial x)_t$.

Though the term $\partial C_1/\partial x$ is independent of the axis to which it is referred, the total flux is not. If the region in which the u_n axis is fixed is moving relative to the x axis with a velocity v, the total flux across the plane $u_n = 0$ is

$$-D_1 \frac{\partial C_1}{\partial u_n} = -D_1 \frac{\partial C_1}{\partial x} \tag{4-2}$$

while the total flux across the same plane when referred to the x axis is

$$-D_1 \frac{\partial C_1}{\partial x} + vC_1 \tag{4-3}$$

If these equations were to be applied to the diffusion of ink in flowing water, Eq. (4-2) gives the flux relative to a chip, while Eq. (4-3) gives the flux of ink relative to an observer on the bank of the stream. Since these equations can apply to either component, e.g., ink or water, equations analogous to Eqs. (4-2) and (4-3) are valid for component 2.

In Chap. 1, it was shown that the conservation of matter leads to the equations:

$$\frac{\partial C_1}{\partial t} = -\frac{\partial J_1}{\partial x} \qquad \frac{\partial C_2}{\partial t} = -\frac{\partial J_2}{\partial x} \tag{4-4}$$

The total number of moles per unit volume (C) is given by the equation $C = C_1 + C_2$. Equations (4-3) and (4-4) for the two components then lead to

$$\frac{\partial C}{\partial t} = \frac{\partial C_1}{\partial t} + \frac{\partial C_2}{\partial t} = \frac{\partial}{\partial x}\left(D_1 \frac{\partial C_1}{\partial x} + D_2 \frac{\partial C_2}{\partial x} - Cv\right) \tag{4-5}$$

† We shall see below that it is most convenient to fix the x axis relative to one end of the diffusion couple, that is, in a region where no diffusion occurs. However, for now it is simply fixed to some arbitrary marker.

If it is assumed that the molar volume of the alloy is independent of composition, C is a constant, and $\partial C/\partial t = 0$. When this is true, Eq. (4-5) can be integrated to obtain

$$D_1 \frac{\partial C_1}{\partial x} + D_2 \frac{\partial C_2}{\partial x} - Cv \equiv I = \text{const} \qquad (4\text{-}6)$$

But in regions far from the initial interface, $\partial C_1/\partial x$ and $\partial C_2/\partial x$ equal zero, so that $-Cv = I$. If we choose to fix x relative to markers placed in such a region, then $v = 0$ in this region and I will equal zero. The equation for the velocity of markers in the diffusion zone relative to the ends of the diffusion couple is thus

$$v = \frac{1}{C}\left(D_1 \frac{\partial C_1}{\partial x} + D_2 \frac{\partial C_2}{\partial x}\right) \qquad (4\text{-}7)$$

Thus we have obtained one equation relating the observable quantity v to the two diffusion coefficients we wish to determine, D_1 and D_2.

To complete the description of the problem, an equation is needed which relates $\tilde{D}$ to D_1 and D_2. This equation can be obtained by substituting Eqs. (4-7) and (4-3) in one of the Eqs. (4-4). This gives

$$\frac{\partial C_1}{\partial t} = \frac{\partial}{\partial x}\left(D_1 \frac{\partial C_1}{\partial x} - \frac{C_1}{C} D_1 \frac{\partial C_1}{\partial x} - \frac{C_1}{C} D_2 \frac{\partial C_2}{\partial x}\right) \qquad (4\text{-}8)$$

From the fact that C is a constant, it follows that $\partial C_1/\partial x = -(\partial C_2/\partial x)$ or

$$\frac{\partial C_1}{\partial t} = \frac{\partial}{\partial x}\left(\frac{C_1 D_2 + C_2 D_1}{C} \frac{\partial C_1}{\partial x}\right) \qquad (4\text{-}9)$$

but this equation is the same as Fick's second law, if we substitute

$$\tilde{D} = \frac{C_1 D_2 + C_2 D_1}{C} = D_1 N_2 + D_2 N_1 \qquad (4\text{-}10)$$

where $N_i = C_i/C$ and $N_1 + N_2 = 1$.
Rearranging Eq. (4-7), we have

$$v = (D_1 - D_2)\frac{\partial N_1}{\partial x} \qquad (4\text{-}11)$$

Equations (4-10) and (4-11) then completely describe the results for the case of isothermal diffusion in an infinite couple. The treatment is independent of the mechanism of diffusion and simply states that if a marker movement is observed, the magnitudes of D_1 and D_2 are different and can be determined at any given composition from measurements of v and $\tilde{D}$ at that composition.

As an example of the use of these equations in determining D_1 and D_2, we shall analyze the results of Smigelskas and Kirkendall. Empirically it was found that the distance between the molybdenum wires placed at the copper-brass interface decreased as $t^{\frac{1}{2}}$. If this is true, then the distance the wires have moved from their initial position (x_m) is given by the equation

$$x_m = \alpha t^{\frac{1}{2}} \tag{4-12}$$

or[1]
$$v = \frac{dx_m}{dt} = \frac{x_m}{2t} \tag{4-13}$$

In this type of diffusion couple any given composition shifts as $t^{\frac{1}{2}}$, so the markers stay at the same composition.[2] The values of D_{Cu} and D_{Zn} can thus be determined for this composition. (In the case of a couple made of copper against 70-30 brass, this composition is found experimentally to be 22.5% Zn.) Smigelskas and Kirkendall found that in 56 days at 785°C each marker shifted 0.0125 cm. Using this and their other data, Darken calculated that at 22.5% Zn

$$D_{Cu} = 2.2 \times 10^{-9} \text{ cm}^2/\text{sec}$$
$$D_{Zn} = 5.1 \times 10^{-9} \text{ cm}^2/\text{sec}$$
$$\frac{D_{Zn}}{D_{Cu}} = 2.3$$

As N_{Zn} approaches zero, Eq. (4-10) indicates that $\tilde{D}$ approaches D_{Zn}. From earlier work by Rhines and Mehl, $\tilde{D} = D_{Zn} = 0.3 \times 10^{-9} \text{ cm}^2/$ sec (at 0% Zn). This indicates that D_{Zn} increases seventeenfold in going from 0 to 22.5% Zn. This is consistent with the magnitude of increase discussed in Chap. 3. However, we will see below that this value of D_{Zn} cannot be equated directly to the diffusion coefficient obtained from a tracer experiment in the absence of a concentration gradient.

Our calculation of D_{Zn} and D_{Cu} in no way acts as a check on the validity of this analysis. To do this we must consider the assumptions leading to Eqs. (4-10) and (4-11). The two most important assumptions made in the analysis were (1) that C is a constant, i.e., that the

---

[1] Equation (4-13) is valid only for markers placed at the initial interface. On the other hand, Eq. (4-11) is valid for a marker no matter where it was initially placed.

[2] This follows from the substitution made on the Boltzmann analysis for an infinite couple (see Sec. 1-5) in which it was shown that $c = f(\lambda) = f(x/2 \sqrt{Dt})$. Thus any c will correspond to a constant value of λ, but a point of constant λ is given by the equation $x = (\text{const}) \sqrt{Dt}$.

molar volume of the alloy does not change with composition, and (2) that the flux relative to the markers is given by $-D_1(\partial C_1/\partial x)$. The first of these is sometimes true, but if it is not, it can be corrected for.[1] Any error in the second one is more difficult to evaluate. To discuss it properly, we must first obtain a more general description of the flux equation.

4-3. PHENOMENOLOGICAL EQUATIONS

In Chap. 1 it was stated that in a binary phase, if the absence of a concentration gradient is an adequate condition for equilibrium, one is safe in using Fick's first law as a flux equation since the flux will go to zero as the system approaches equilibrium. This is applicable to many systems, and since the concentration is easily measured, it is commonly used. However, $\partial c/\partial x = 0$ is a very restricted condition for equilibrium. To gain insight into just what the limitations of this condition are, it is necessary to use a more general condition for equilibrium.

To obtain this general equation, one proceeds as follows. For a given n-component system at equilibrium, the system can be uniquely determined by specifying T, P, μ_1, μ_2, . . . , μ_{n-1}, and ϕ, where μ_i is the chemical potential[2] and ϕ is any relevant scalar potential, e.g., electric potential. If now the system is displaced slightly from equilibrium, it seems most likely, and is certainly simplest, to assume that the rate of return to equilibrium is proportional to the deviation from equilibrium. And, until it is proved to be unnecessary, the flux of, say, component 1 is assumed to be proportional to the gradient of each of the potentials listed above. Thus the most general equation for J_1 is

$$J_1 = -M_{11}\frac{d\mu_1}{dx} - M_{12}\frac{d\mu_2}{dx} - \cdot \cdot \cdot - M_{1n}\frac{d\mu_n}{dx} - M_{1t}\frac{dT}{dx}$$
$$- M_{1P}\frac{dP}{dx} - M_{1\phi}\frac{d\phi}{dx} \quad (4\text{-}14)$$

---

[1] R. W. Balluffi, *Acta Met.*, **8**: 871 (1960).

[2] The chemical potential μ_i is defined by the equation

$$\mu_i = \left(\frac{\partial G}{\partial n_i}\right)_{P,T,n_j} \quad i \neq j$$

where G is the Gibbs free energy of the subsystem or phase. If the free energy of the total system is to be minimized at constant P and T, this is equivalent to the requirement that μ_i be the same in all parts of the system.

The equations for the flux of component 1, the flux of heat, etc., are given by comparable sums. These are called the phenomenological equations since they stem from no model, but from the observed conditions of equilibrium. A general discussion of the justification of these equations, as well as their simplification and manipulation can be found in de Groot.[1] Experimentally it is found that M_{1T}, $M_{1\phi}$, and M_{1P} *are not zero*, but for now we are interested only in the fact that in an isothermal, isobaric, isopotential system, J_1 is not only proportional to $d\mu_1/dx$, but may be proportional to $d\mu_2/dx$, $d\mu_3/dx$, etc., as well. A complete, concise discussion of the application of the phenomenological equations to alloy diffusion is given by Bardeen and Herring.[2] We shall discuss here only the assumptions required to obtain the flux equation assumed by Darken.

Consider a one-dimensional diffusion problem in a two-component system. If we allow the possibility of a flux of vacancies (J_v) in addition to the fluxes J_1 and J_2,† the flux equations are given by

$$J_1 = -M_{11}\frac{d\mu_1}{dx} - M_{12}\frac{d\mu_2}{dx} - M_{1v}\frac{d\mu_v}{dx} \tag{4-15}$$

$$J_2 = -M_{21}\frac{d\mu_1}{dx} - M_{22}\frac{d\mu_2}{dx} - M_{2v}\frac{d\mu_v}{dx} \tag{4-16}$$

$$J_v = -M_{v1}\frac{d\mu_1}{dx} - M_{v2}\frac{d\mu_2}{dx} - M_{vv}\frac{d\mu_v}{dx} \tag{4-17}$$

In any region where lattice sites are neither created nor destroyed, the three fluxes are related by the equation

$$J_1 + J_2 + J_v = 0 \tag{4-18}$$

If this is to be true for any value of each of the gradients, substitution of Eqs. (4-15) to (4-17) into (4-18) shows that we must have

$$M_{11} + M_{21} + M_{v1} = 0 \qquad M_{12} + M_{22} + M_{v2} = 0$$
and
$$M_{1v} + M_{2v} + M_{vv} = 0$$

For the equations written here, there is a set of reciprocity relations due to Onsager which state that $M_{ij} = M_{ji}$, so that $M_{12} = M_{21}$,

---

[1] S. R. de Groot, "Thermodynamics of Irreversible Processes," Interscience Publishers, Inc., New York, 1952. A more readable but less complete treatment is given by K. G. Denbigh, "The Thermodynamics of the Steady State," Methuen & Co., Ltd., London, 1950.

[2] J. Bardeen and C. Herring, "Atom Movements," p. 87, ASM, Cleveland, 1951.

† This does not assume that a vacancy mechanism is dominant; it only allows the possibility. If a vacancy mechanism does not operate, then $J_v = 0$.

$M_{v1} = M_{1v}$, and $M_{v2} = M_{2v}$. Using all the equations between the M_{ij} gives

$$J_1 = -M_{11}\frac{d}{dx}(\mu_1 - \mu_v) - M_{12}\frac{d}{dx}(\mu_2 - \mu_v)$$

$$J_2 = -M_{21}\frac{d}{dx}(\mu_1 - \mu_v) - M_{22}\frac{d}{dx}(\mu_2 - \mu_v)$$

(4-19)

These are now the simplest equations which apply in general for a binary alloy where a vacancy mechanism is allowed. If a vacancy mechanism does not operate, the vacancy concentration will be at equilibrium at all points, and $d\mu_v/dx$ will equal zero (as will μ_v).

To obtain Darken's flux equation, two additional assumptions must be made. These are (1) that the vacancies are everywhere in thermal equilibrium, or $\mu_v \simeq 0$, and (2) that the off-diagonal coefficients M_{12} and M_{21} are essentially zero. Inserting these two assumptions in the first of Eq. (4-19) gives

$$J_1 = -M_{11}\frac{d\mu_1}{dx}$$

(4-20)

To relate D_1 to M_{11} in Eq. (4-20) we equate our two expressions for J_1.

$$J_1 = -M_{11}\frac{d\mu_1}{dx} = -D_1\frac{dC_1}{dx}$$

(4-21)

In Sec. 1-5, other equations were given for this flux. If the force on an atom (F) was taken to be the gradient of the appropriate potential and v was the mean velocity of the atom when acted upon by F, the mobility B was given by the equation $B = v/F$. Combining the flux equations used in Sec. 1-5 with Eq. (4-21) gives[1]

$$J_1 = C_1 v = B_1 F_1 C_1 = -B_1 C_1\frac{d\mu_1}{dx} = -M_{11}\frac{d\mu_1}{dx} = -D_1\frac{dC_1}{dx}$$

(4-22)

It is apparent that $M_{11} = B_1 C_1$ and that

$$D_1 = B_1\frac{d\mu_1}{d \ln C_1} = B_1\frac{d\mu_1}{d \ln N_1}$$

(4-23)

[The second equality in Eq. (4-23) follows since if $C_1/C = N_1$ then

---

[1] Here the equation $F = -d\mu_1/dx$ replaces the equation $F = -dV/dx$ used in Sec. 1-5.   The concentration gradient that enters $d\mu_1/dx$ does not exert mechanical force in the sense that a potential-energy gradient does, but it does produce a net flux of atoms and thus can be thought of as a force.   Also one could take the viewpoint that this is required if the flux given by Eq. (4-22) is to equal zero at equilibrium.

$d \ln C_1 = d \ln N_1$.] The equation relating μ_i and N_1 is

$$\mu_1 = \mu_0(T,P) + RT(\ln N_1 + \ln \gamma_1)$$

where γ_1 is called the activity coefficient of 1. Thus

$$\frac{d\mu_1}{d \ln N_1} = RT\left(1 + \frac{d \ln \gamma_1}{d \ln N_1}\right)$$

and

$$D_1 = B_1 RT\left(1 + \frac{d \ln \gamma_1}{d \ln N_1}\right) \tag{4-24}$$

The same sort of relation holds for component 2. In dilute solution, γ_1 is a constant. (This follows from Raoult's law for the solvent and from Henry's law for the solute.) Thus in dilute or ideal solution, $D_1 = B_1 RT$, but in concentrated, nonideal solutions the ratio $D_1/B_1 RT$ will differ from unity. The sign and magnitude of the deviation will depend on the type and degree of nonideality.

4-4. RELATIONSHIP BETWEEN CHEMICAL D_1 AND TRACER D_1^*

Consider a binary diffusion couple in which there is no chemical concentration gradient, that is, $dN_1/dx = 0$, but in which there is an isotopic concentration gradient, that is, $dN_1^*/dx \neq 0$. (N_1^* is the mole fraction of radioactive component 1.) Using Eq. (4-24), the self-diffusion coefficient in an alloy is given by the equation

$$D_1^* = B_1^* RT\left(1 + \frac{d \ln \gamma_1^*}{d \ln N_1^*}\right)_{N_1+N_1^*} \tag{4-25}$$

But the stable and the radioactive isotopes are assumed to be chemically identical in such an experiment, so γ_1 should depend only on the overall value of $N_1 + N_1^*$ and not on the relative proportions of N_1^* and N_1. Stated in another manner, for $N_1^* + N_1$ constant, $(d \ln \gamma_1^*/d \ln N_1^*)$ should equal zero. Thus

$$D_1^* = B_1^* RT \tag{4-26}$$

From this same chemical identity of the two isotopes, it seems reasonable to assume that the mobilities determined in the two types of experiments are the same, that is, $B_1^* = B_1$.† This leads to the desired

---

† Bardeen and Herring, *loc. cit.*, show that, for a pure metal $B_1 f = B_1^*$, where f is the correlation coefficient (see also footnote 2, p. 142). Thus we are here assuming something about the relative contribution of correlation effects to experiments in which B_1 and B_1^* are determined. Our assumption is related to taking $M_{ij} = 0$ ($i \neq j$) in Eq. (4-19). This will be discussed further in the next section.

equation

$$D_1 = D_1^* \left(1 + \frac{d \ln \gamma_1}{d \ln N_1}\right) \tag{4-27}$$

That is, the diffusion coefficient for component 1 in a chemical concentration gradient D_1 is not equal to the value D_1^* obtained in a self-diffusion experiment except in ideal or dilute solutions. In concentrated nonideal solutions D_1 and D_1^* will differ. This difference arises from the fact that with a given gradient dN_1/dx or dN_1^*/dx, the actual driving force, that is, $d\mu_1/dx$, depends upon whether there is a variation in $N_1 + N_1^*$ along the diffusion direction or simply a variation of N_1/N_1^* with $\nabla(N_1 + N_1^*) = 0$.

The most common type of diffusion data is for $\tilde{D}$, D_1^*, and D_2^*. Thus an experimental check of this analysis can be made by relating these three quantities. From the Gibbs-Duhem equation

$$N_1 d\mu_1 + N_2 d\mu_2 = 0$$

and from the definition of μ_i as a function of N_i

$$N_i d\mu_i = RT(dN_i + N_i d \ln \gamma_i)$$

Substitution of the latter into the former and the fact $dN_1 = -dN_2$ gives

$$N_1 \frac{d \ln \gamma_1}{dN_1} = N_2 \frac{d \ln \gamma_2}{dN_2}$$

Combining this with Eqs. (4-10) and (4-27) gives

$$\tilde{D} = (D_1^* N_2 + D_2^* N_1)\left(1 + \frac{d \ln \gamma_1}{d \ln N_1}\right) \tag{4-28}$$

All the terms appearing in this equation can be measured, so a check on the analysis can be made by comparing the calculated and observed values of $\tilde{D}$. Using Johnson's data on a 50:50 silver-gold alloy at 1000°C gives[1]

$$\tilde{D} = 9.7 \times 10^{-9} \text{ cm}^2/\text{sec}$$
$$D_{Ag}^* = 8.2 \times 10^{-9} \text{ cm}^2/\text{sec}$$
$$D_{Au}^* = 3.2 \times 10^{-9} \text{ cm}^2/\text{sec}$$

Using the self-diffusion data and available thermodynamic data, Darken calculated $\tilde{D} = 9.3 \times 10^{-9}$ cm²/sec. In view of the uncertainty in the thermodynamic data and the lesser uncertainty in the

[1] W. A. Johnson, *Trans. AIME*, **142**: 331 (1947).

diffusivities, the agreement between the calculated and experimental values of $\tilde{D}$ is well within the experimental error. More recent studies have also borne out the validity of Eq. (4-28).[1] The agreement between the calculated and experimental results is usually within the experimental error, though the comparison of the two values of $\tilde{D}$ cannot be made with precision because of the size of the errors in measuring $\tilde{D}$ and ($d \ln \gamma_i/d \ln N_i$).

4-5. TEST OF DARKEN'S ASSUMPTIONS

It was seen above that Darken's relation between $\tilde{D}$, D^*_{Ag}, and D^*_{Au} was borne out by the data. This in itself is strong evidence that the analysis is basically correct. However, if we are to develop a sense of the generality of this validity, it is helpful to have more of a mechanistic feeling for the assumptions $\mu_v \simeq 0$, $B_i = B^*_i$, and $M_{ij} = 0$ $(i \neq j)$.

The Assumption $\mu_v \simeq 0$. This assumption requires that the concentration of vacancies be maintained at its equilibrium value at every point in the diffusion couple. The interstitial mechanism of diffusion in fcc metals was never seriously considered, and since the Kirkendall effect showed that $D_1 \neq D_2$, a ring mechanism was untenable. Thus diffusion appears to occur by a vacancy mechanism. If this is true, a volume of vacancies equal to the volume swept out by the markers, i.e., marker shift times cross-sectional area of the couple, must be created on one side and destroyed on the other. The problem of understanding the assumption $\mu_v \simeq 0$ is then one of determining how and where these vacancies are formed and destroyed. The areas of formation and removal can be determined by operating on the experimental $C(x)$ curve. Figure 4-2a shows an assumed $C_1(x)$ curve for the case of $D_1 < D_2$. In (b) the fluxes have been obtained by taking the gradient of $C_1(x)$ and multiplying by D_1 and D_2, respectively. The vacancy flux is equal to the difference between J_1 and J_2 and is in the same direction as J_1. In (c) the divergence of J_v is shown, that is, the rate of generation ($dJ_v/dx > 0$) or destruction ($dJ_v/dx < 0$). From this figure it is seen that most of the vacancies are produced near one end of the diffusion zone and are removed near the other.

The first suggestion was that the surface served as sources and sinks for the vacancies, but when the marker movement was independent of specimen size, the grain boundaries were suggested. When the grain size was shown to have no effect on the shift, workers were forced to conclude that there were many potential sources and sinks within a

[1] Reynolds, B. Averbach, and M. Cohen, *Acta Met.*, **5**: 29 (1957). J. Manning, *Phys. Rev.*, **116**: 69 (1959).

given grain.   An edge dislocation would serve the purpose, but it would be grown out of the grain in absorbing or giving off one plane of vacancies.   A longer-lived source is obtained by having a screw dislocation parallel to the diffusion direction which rotates into an edge dislocation in a plane normal to the flux.   This edge dislocation can then rotate around the screw dislocation, giving off or taking on a plane

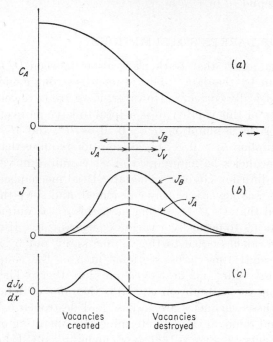

Fig. 4-2. (a) Assumed concentration-distance curve for component A. (b) Fluxes of A and B that result $(D_B > D_A)$. The flux of vacancies will be equal to the difference between J_A and J_B. (c) dJ_v/dx equals the rate of creation of vacancies at that point.

of vacancies in each revolution.   The geometry is quite similar to that given by Burton, Cabrera, and Frank[1] in their discussions of crystal growth from the vapor.   In our case, their ledge is replaced by an internal edge dislocation.   Bardeen and Herring[2] give calculations to show that the supersaturation S required to operate a source or sink of

[1] Burton, N. Cabrera, and F. Frank, *Phil. Trans. Roy. Soc. London*, **243A**: 299 (1951).

[2] Bardeen and Herring, "Atom Movements," ASM, Cleveland, 1951, but see also D. Kuhlmann-Wilsdorf, R. Maddin, and H. Wilsdorf, "Strengthening Mechanisms in Solids," ASM, Cleveland, 1961.

this type is about 0.01, where

$$S \equiv \frac{N_v \text{ (actual)}}{N_v \text{ (equilib)}} - 1 \qquad (4\text{-}29)$$

From the phenomenological viewpoint, one can say that

$$\frac{dJ_v}{dx} \approx -\mu_v \approx S \qquad (4\text{-}30)$$

That is, the rate of creation or removal of vacancies is proportional to the deviation of μ_v from its equilibrium value of zero. Thus the line shown in Fig. 4-2c could also be taken as $-\mu_v$. This variation of μ_v with distance tends to decrease the flux J_2 and increase J_1 relative to the values predicted by Darken's equations [see Eqs. (4-19) and (4-20)]. Any deviation of μ_v from zero will therefore tend to make the observed value of D_1/D_2 closer to unity than it should be.

Two types of experiments have been performed to determine whether the nonequilibrium vacancy concentration would have a measurable effect on the observed values of D_1 and D_2. The first experiment measured D_1 and D_2 in concentration gradients of varying magnitude.[1] The second estimated the maximum value of $\partial \mu_v/\partial x$ from an analysis of pore formation and compared it with the maximum value of $\partial \mu_1/\partial x$.† In both cases, it was concluded that any error caused by $\mu_v \neq 0$ was not detectable in the experiments performed.

The Assumptions $B_i = B_i^*$ and $M_{ij} = 0$. Darken's equations agree with the experimental results to within an appreciable experimental error. This shows that they provide a valid approximation to physical reality, but they are only a first approximation. The approximation is made when B_1 is taken equal to B_1^*. This can best be shown by emphasizing the differences between the two separate experiments in which B_1 and B_1^* (or D_1 and D_1^*) are measured. B_1^* is determined from the rate at which stable and radioactive isotopes of component 1 become intermingled in a chemically homogeneous alloy; in the tracer experiment there is neither a net flux of matter nor a net flux of vacancies. B_1 is determined from the rate at which stable atoms flow down a chemical concentration gradient; there is both a net flux of matter and a net flux of vacancies in this experiment. Manning has given a detailed analysis of the relation between these experiments and finds that the difference between B_1^* and B_1 is proportional to the net

---

[1] G. Horne and R. Mehl, *Trans. AIME*, **203**: 88 (1955).

† R. Balluffi, *Acta Met.*, **2**: 194 (1954); R. Balluffi and L. Siegle, *Acta Met.*, **5**: 449 (1957).

flux of vacancies present in a Kirkendall experiment.[1]   This flow of vacancies means that in addition to their random motion, the vacancies drift in a direction opposite to that of the flux of the faster moving component.   Thus vacancies will more frequently approach any given atom from one side than from the other.   As a result, the flux of either component is that given by the random motion of the vacancies, plus an additional part due to the vacancy flow.   The part due to the random motion of vacancies is given by assumption $B_1 = B_1^*$; the part due to the vacancy flow gives an additional contribution.   As a result of this contribution, a second term, proportional to the net vacancy flow, is added to Eq. (4-28) and a comparable equation relating the marker movement x_m to D_1^* and D_2^*.   Manning estimates that the contribution of these terms is large enough to be measurable, but to date no adequate data are available.

Some of this effect of vacancy flow can be included in a modified relation between D_1 and D_1^*, and part of it will appear as a nonzero value of M_{ij}.   The proper allocation of this effect between the two is a difficult, and partly arbitrary, decision.   The interested reader is referred to Manning's paper.

4-6. TERNARY ALLOYS

By now the increase in complexity that can arise in going from diffusion in pure metals to binary systems should be established in the reader's mind.   It is not difficult then to believe that diffusion in ternary systems can be even more complicated.   This type of system has not been extensively studied, nor is it well understood.   We will describe here only one experiment on ternaries.   This indicates a few of the new problems that arise in going from a binary to a ternary system.   It is also relevant to the problem, mentioned earlier, of the correct flux equation to be used in treating diffusion.

Consider the following experiment.   A bar of an Fe–0.4% C alloy is joined to an Fe–0.4% C–4% Si alloy and diffused at 1050°C where the couple consists of only one phase (fcc austenite).   Since there is no carbon concentration gradient, Fick's first law would predict no flux of carbon.   Darken[2] has performed essentially this experiment and obtained the carbon distribution shown in Fig. 4-3.   It is seen that carbon diffusion has increased the concentration gradient in the couple.

It has been shown in other studies of Fe-Si-C alloys that adding

[1] J. R. Manning, *Phys. Rev.*, **124**: 470 (1961).
[2] L. Darken, *Trans. AIME*, **180**: 430 (1949).

silicon to Fe-C alloys increases the chemical potential of carbon.   Thus upon annealing, the rapidly diffusing carbon is redistributed in the region of the joint to give local equilibrium, that is, to eliminate the gradients in the chemical potential of carbon.   Since there is a very rapid change in the concentration of silicon near the joint, this local equilibrium requires an equally rapid change in the carbon concentration.   Using the results of the prior thermodynamic studies, Darken showed that the chemical potential of carbon was indeed constant through the joint.   Thus the results could have been correctly and easily described with a flux equation of the form $J = -M(\partial\mu/\partial x)$, while the application of Fick's first law was virtually impossible.

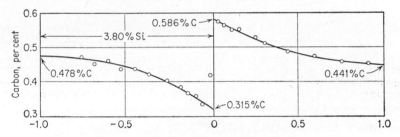

FIG. 4-3. Distribution of carbon resulting from 13-day anneal at 1050°C. The carbon content was initially 0.478% throughout the left side and 0.441% throughout the right side.   (*From L. S. Darken, "Atom Movements," p. 21, ASM, Cleveland, 1951.*)

This furnishes a striking example of the effects which can arise in ternary diffusion couples.   The effect is exaggerated by the very rapid diffusion of the interstitial carbon relative to that of silicon.   However, in a ternary alloy with two substitutional solutes a flux is possible in the absence of a concentration gradient or even against a concentration gradient.   In these cases the values of D obtained using Fick's laws are empirical constants which depend on the gradient of composition as well as the composition.   As a result, they are of little practical value and no theoretical value.

Before closing our discussion of Darken's experiment, it is interesting to trace out the changes in composition at two points on opposite sides of the couple using a ternary diagram.   Figure 4-4 schematically shows such a trace.   Note that the points initially move along lines of constant silicon concentration.   This is another way of indicating that the carbon diffuses much more rapidly than the silicon.   If the silicon and the carbon diffused at about the same rate, much of this curvature would be absent.

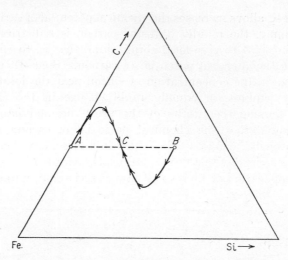

FIG. 4-4. Schematic diagram showing the change in composition of two points on opposite sides of the weld in Darken's diffusion couple of Fe–0.44% C and Fe–0.48% C, 3.8% Si. [*From L. S. Darken, Trans. AIME*, **185**: 435 (1959).]

4-7. DIFFUSION IN MULTIPHASE BINARY SYSTEMS

A problem of considerable practical importance has to do with the rate of growth of one phase in or on another, where the rate of growth is diffusion controlled. Parabolic oxidation is the best-studied example of this,[1] although there are many others. The only principle involved that has not been discussed before is the rule of two-phase equilibrium, i.e., when two phases are in equilibrium, the chemical potential of each component will be the same in each phase. The importance of this principle to the problem can be seen by considering Fig. 4-5. Here a protective oxide is shown between gaseous oxygen and a metal. Figure 4-5b shows the variation of the oxygen concentration with position, while 4-5c shows the variation of the chemical potential of the oxygen with position.[2] The important point is that the chemical potential is taken to be continuous across the two-phase interfaces. This means that the two adjoining phases are taken to be in equilibrium *at the interfaces*. This is usually the case since the attainment

[1] O. Kubaschewski and B. Hopkins, "Oxidation of Metals and Alloys," Academic Press, Inc., New York, 1953. K. Hauffe, in "Progress in Metal Physics," vol. 4, pp. 71–104, Pergamon Press, Inc., New York, 1953.

[2] A lucid qualitative discussion of the relation between the phase diagram and the phases formed upon combining two pure components is given by F. N. Rhines, "Phase Diagrams," pp. 50, 68, 107, McGraw-Hill Book Company, Inc., New York, 1958.

of this local equilibrium at the interfaces requires diffusion of only a few angstroms distance, while the draining of the components from the region of the interface requires diffusion over much longer distances.

If this local equilibrium exists, the concentration at the boundaries of each phase is known (from the phase diagram), and if $\tilde{D}(c)$ is known, the diffusion problem is mathematically determined. The mathematical problem is not a simple one since it involves a variable $\tilde{D}$ and moving boundaries. Nevertheless, it is well defined and not of real interest here.[1]

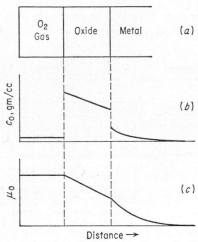

FIG. 4-5. (a) Schematic drawing of system in which a protective oxide is formed on a piece of metal which is being oxidized. (b) A plot of the oxygen concentration vs. distance. (c) A plot of the oxygen chemical potential vs. distance.

Most of the fundamental diffusion studies in this area concern the oxidation of metals. Here the primary problem is the determination of the rate-determining diffusion process and the ways in which this can be effected. This problem will be discussed in the next chapter.

4-8. VARIATION OF $\tilde{D}$ ACROSS A BINARY PHASE DIAGRAM

The metallurgist is often faced with the problem of making "reasonable approximations" for systems or alloys in which no accurate measurements have been made. In the particular case of diffusion he may be given the problem of estimating the relative value of $\tilde{D}$ across much of a phase diagram which contains several intermediate phases.

---

[1] The reader interested in solutions to this problem is referred to W. Jost, "Diffusion," chap. 1, Academic Press, Inc., New York, 1951.

$\tilde{D}$ has been measured in a few such alloy systems, and several helpful generalizations can be made. Two of these follow.

1. If adding A to B lowers the melting point of B, or the liquidus line, it will also increase $\tilde{D}$ at any given temperature. If A raises the melting point of B, $\tilde{D}$ will decrease. This rule is a variation of the observation that for a given crystal structure, $\tilde{D}$ at the liquidus is roughly a constant or $\Delta H/T_{\text{liq}}$ is roughly a constant. As an example, see Fig. 4-6.

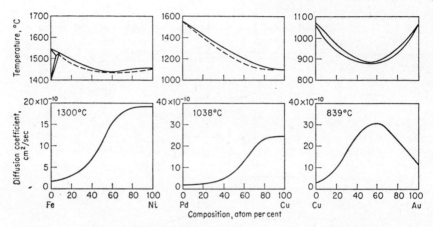

FIG. 4-6. Phase diagrams and the variation of $\tilde{D}$ with composition for the iso-morphous systems iron-nickel, copper-palladium, and copper-gold. (*From C. E. Birchenall, "Atom Movements," p. 122, ASM, Cleveland, 1951.*)

2. For a given solvent, at a given temperature and composition, diffusion will be much faster in a bcc lattice than in a close-packed lattice. This is true for both interstitial and substitutional solutes.

As examples of the second point, consider the following order of magnitude ratios. For carbon in essentially pure iron at 910°C, $D(\alpha)/D(\gamma) \simeq 10^2$.† For iron in iron at 850°C, $D(\alpha)/D(\gamma) \simeq 10^2$.‡ For the comparison of a bcc and an hcp lattice, in zirconium at 825°C, $D(\beta)/D(\alpha) \simeq 10.$§ The more rapid diffusion in the bcc modification

† R. P. Smith, *Acta Met.*, **1**: 578–587 (1953).   C. Wert, *Phys. Rev.*, **79**: 601 (1950).

‡ C. Birchenall and R. Mehl, *Trans. AIME*, **188**: 144 (1950).

§ E. V. Borisov et al., "Technical Sciences and Industrial Uses of Isotopes," All-Union Conference on Application of Radioactive and Stable Isotopes and Radiation in the National Economy and Science, Moscow, Apr. 2–5, 1957, p. 31, Consultants Bureau, Inc., New York, 1958.

is usually "explained" by saying that the bcc lattice is more loosely packed or a more open structure. This is qualitatively consistent with our earlier discussion of the ion core repulsion to the activation energy. No quantitative, theoretical rationalization has been made.

A more detailed discussion of this kind of qualitative correlation has been given by Birchenall.[1]

PROBLEMS

4-1. Two markers are placed in a couple formed of two semi-infinite regions of A and B—one at the initial weld interface and one a short distance away. Show qualitatively how the position of each varies with time if $D_A > D_B$. Derive these curves by plotting $N(x)$ and $\partial N/\partial x$ versus x and using the equation

$$v = (D_A - D_B)\, \partial N_A/\partial x.$$

4-2. It is frequently stated that if a reaction or mixing is diffusion controlled, any given composition, e.g., an α-β interface, will move as $\sqrt{t}$ even when D varies with composition. Show under what conditions this is true for the case of $D = f(c)$. (It may be helpful to first assume $D = $ const and show that the above is true for one general set of boundary conditions but is not true for another general set.)

4-3. Using Eqs. (4-26) and (4-27), calculate $d \ln D/d(1/T) \equiv -\Delta H/R$ for D_1 and D_1^*. Discuss the reason for, and the sign of, the difference in the two.

4-4. Darken derived an equation relating the diffusion coefficients of component 1 (in a binary alloy) for the two cases of diffusion in the presence of chemical concentration gradient D_1 and diffusion in a homogeneous alloy with a gradient only in the relative concentration of the isotopes of component 1, D_1^*.

(*a*) Give Darken's equation for this case.

(*b*) State in 50 words or less why these two diffusion coefficients differ (no equations).

(*c*) Would the apparent activation energy for D_1 equal that for D_1^*? [For a discussion of 4-4c, see J. Hilliard's Appendix to paper by Reynolds, Averbach, and Cohen, *Acta Met.*, **5**: 29 (1957)].

4-5. Figure 4-7 shows the reported variation of penetration depth with time for Armco Iron carburized in a graphite-BaCO$_3$ mixture at 1830°F with, and without, ultrasonic vibration [*Metal Prog.*, **79**: 79–83 (1961).] This increase in penetration depth could result from the ultrasonic vibration either (1) increasing the diffusion coefficient or (2) changing the boundary conditions. Two experimental procedures which will distinguish between these two causes are given by the answers to the following. (Assume that D does not vary with composition.)

(*a*) For the case in which only (1) is true, plot two $c(x)$ curves on the same graph, one for a specimen with and one for a specimen without ultrasonic vibration (take d to be the depth at which c has decreased to one-third its surface value).

[1] C. E. Birchenall, "Atom Movements," p. 112, ASM, Cleveland, See also D. Sherby and M. Simnad, *Trans. ASM*, **54**: 227 (1961).

(b) For the case in which only (2) is true, plot the $c(x)$ curves for specimens with and without ultrasonic vibration.

(c) Show how $\partial c/\partial x$ at $x = 0$ varies with time if $c(x = 0)$ does not change with time.

(d) Show how $(\partial c/\partial x)_{x=0}$ changes with t if $c(x = 0)$ is determined by some reaction which furnishes carbon to the surface at a constant rate (a possibility for the case with no ultrasonic vibration).

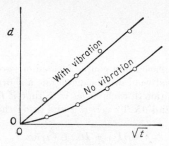

FIG. 4-7

4-6. Chromium diffuses into pure iron at 1000°C under conditions which maintain a concentration of 50% chromium at the surface. The Fe-Cr system has a γ loop; at 1000°C, the maximum Cr content of the γ-phase is 12%, and the minimum Cr content of the α-phase is 13%. The diffusivity D of Cr is greater in the α-phase than in the γ-phase. Sketch a schematic diagram of concentration of solute vs. depth below the surface. Pay special attention to slopes and curvatures. Do not draw to scale but indicate absolute values where possible.

4-7. In a given phase the flux of a given component is always in the direction of its decreasing chemical potential. Therefore it is helpful to establish the following. Using the thermodynamic relation between μ_1 and N_1 show that

(a) For a binary system if $dN_1/dx = 0$, then $d\mu_1/dx = 0$, and if $d\mu_1/dx > 0$, then $dN_1/dx > 0$. (x is distance.)

(b) For a ternary system, if $d\mu_1/dx > 0$, dN_1/dx can be either >0 or <0.

4-8. The simplest flux equation for diffusion in a ternary, which uses the chemical potential, is

$$J_1 = -M_1 \frac{\partial \mu_1}{\partial x}$$

If $\mu_1 = \text{const} + RT \ln \gamma N_1$ (where γ varies with both N_1 and N_2), show that

$$D_1 \equiv -\frac{J_1}{dN_1/dx} = M_1 RT \left(\frac{1}{N_1} + \frac{\partial \ln \gamma}{\partial N_1} + \frac{\partial \ln \gamma}{\partial N_2} \frac{dN_2}{dN_1} \right)$$

If dN_2/dN_1 is rewritten $(dN_2/dx)/(dN_1/dx)$, discuss how, at a given composition, this quantity can vary with the type of couple used. (A ternary diagram is helpful in showing the various values of this ratio.)

4-9. Explain how the D_1 and D_1^* in Darken's analysis are determined experimentally, and tell why they differ.

chapter 5 DIFFUSION IN
NONMETALS

In the preceding chapters, the specific examples used concerned metals. This stems partly from the author's experience but also from the fact that the majority of the research on diffusion in solids has been done with metals. There is reason to believe that all the general theory and most of the physical phenomena discussed in the earlier chapters applies equally well to nonmetals, although well-studied examples are not available in many cases. Because of their special electrical properties, several new effects are found in ionic solids and semiconductors which are not found in metals. In this chapter we shall deal only with the phenomena which are unique to nonmetals.

The majority of the chapter deals with the relation between electrical conductivity and diffusion in ionic solids. This topic has been studied extensively and successfully, and a relatively complete picture of the subject can be given. The latter part of the chapter deals with semiconducting materials. Here the mobility of the electrons destroys the simple relation between the conductivity and the diffusion coefficient. However, the defect concentration is still simply related to the composition, and some interesting and novel effects can be demonstrated in this way. The chapter closes with a brief discussion of diffusion in ordered alloys.

5-1. DEFECTS IN IONIC SOLIDS

The reader may recall that the forces between atoms in an ionic crystal are largely classical and that a well-developed theory of ionic

137

crystals was established before the advent of quantum mechanics. The physical model resulting from such studies is demonstrated by the NaCl-type lattice shown in Fig. 5-1.   Here the equally charged ions are arranged so that the oppositely charged ions are nearer to each other and the similarly charged ions are farther apart.   If diffusion occurred by the interchange of a neighboring sodium ion and a chlorine ion, these ions would go to sites which were surrounded by ions of the same charge.   The increase in electrostatic energy of this new configuration over the normal situation is so great that diffusion by this mechanism is

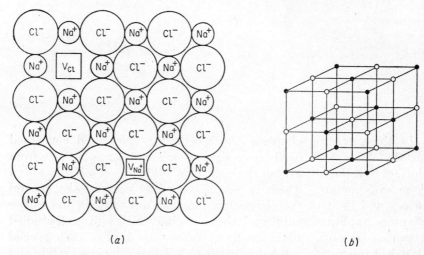

(a) (b)

FIG. 5-1. (a) Schematic drawing of (100) plane of the NaCl lattice showing the relative sizes of the ions, a sodium vacancy $\boxed{V_{\text{Na}}}$, and a chlorine vacancy $\boxed{V_{\text{Cl}}}$. (b) Spacial arrangement of the ions in the NaCl unit cell.

out of the question.   Calculation of the energy required to form various mobile defects indicates that the dominant defects are always vacancies and interstitials.

Since there are two types of ions in these compounds, the formation of defects is not as simple as in the case of a pure metal.   For example, if metal ion (cation) vacancies were formed at a surface of an ionic solid and then diffused into the crystal, the surface would have an excess positive charge; the inside of the crystal would have an equal negative charge.   These separated, unlike charges would have a very large electrostatic energy per vacancy, so large in fact that the separation of unlike charges over macroscopic distances does not occur.   To have defects which will maintain local charge neutrality, two kinds of defects of opposite charge must be formed.   For example, if both an anion vacancy and a cation vacancy were formed, charge neutrality would be preserved. When an equal number of anion and cation vacancies are formed, the resulting disorder is said to be of the Schottky type. This type of disorder is the dominant type of disorder in alkali halide crystals and is shown in Fig. 5-1a.

An equation for the equilibrium fraction of anion sites that are vacant (N_{va}) and the equilibrium fraction of cation sites that are vacant (N_{vc}) can be obtained in a manner analogous to that used in Chap. 2 to calculate N_v. If δn anion vacancies and δn cation vacancies are added to a crystal, and if all of the vacancies are randomly distributed, the change in free energy of the system will be

$$\delta G = \frac{\delta n}{N} \left[\Delta H_{va} + \Delta H_{vc} - T(\Delta S_{va} + \Delta S_{vc}) + RT(\ln N_{va} + \ln N_{vc}) \right]$$

$$(5\text{-}1)$$

Where $-R \ln N_{va}$ is the ideal entropy increase for mixing anion vacancies into the crystal, ΔS_{va} is the molar entropy of formation of anion vacancies, and ΔH_{va} is the molar enthalpy of formation of anion vacancies. Similar definitions apply to $-R \ln N_{vc}$, ΔS_{vc}, and ΔH_{vc}; N is Avogadro's number. Setting $\Delta G_{va} = \Delta H_{va} - T\Delta S_{va}$ and $\Delta G_{vc} = \Delta H_{vc} - T\Delta S_{vc}$, the condition for equilibrium is

$$(N_{va})(N_{vc}) = \exp \frac{-\Delta G_{va} - \Delta G_{vc}}{RT} = \exp\left(-\frac{\Delta G_S}{RT}\right) \qquad (5\text{-}2)$$

where ΔG_S is the molar free energy of formation of the pair of vacancies.

If the free energy required to form an interstitial cation (ΔG_{ic}) is much less than that required to form an anion vacancy (ΔG_{va}), the charge of the cation vacancies will be compensated by metal ions going into interstitial sites. This combination of defects is known as Frenkel disorder. If N_{ic} is the fraction of interstitial sites which is occupied by metal ions (cations) and the defects are all randomly distributed, the equilibrium condition is

$$(N_{ic})(N_{vc}) = \exp\left(-\frac{\Delta G_F}{RT}\right) \qquad (5\text{-}3)$$

where ΔG_F is the molar free energy of formation for a pair of Frenkel defects, i.e., an interstitial plus a vacancy. This type of disorder is dominant in AgCl and AgBr.

It should be emphasized that in our derivations it was not necessary to assume that $N_{vc} = N_{va}$ [in Eq. (5-2)] or that $N_{ic} = N_{vc}$ [in Eq. (5-3)]. These two equations are thus analogous to the mass action constants found in discussions of chemical equilibrium. Two cases commonly arise in which these general properties of the equations are used.   First, if $\Delta G_{ic} \simeq \Delta G_{va}$, then $\Delta G_S \simeq \Delta G_F$; and, in addition to cation vacancies, both anion vacancies and cation interstitials will be present.   The relative concentrations of the various defects must then be determined by simultaneously satisfying Eqs. (5-2) and (5-3) and the condition of charge neutrality which requires that $N_{vc} = N_{va} + N_{ic}$ (provided all ions have the same charge).   A second problem, which will be developed in detail below, arises if some of the matrix ions are replaced by ions of a different valence.   For example, if Ca^{++} ions replace a few Na^{+} ions in NaCl, a proportionate number of defects with a net negative charge must be added to maintain charge neutrality. Thus the equilibrium concentration of defects will be determined by the arrangement which will maintain charge neutrality and at the same time satisfy Eqs. (5-2) or (5-3).

5-2. DIFFUSION AND IONIC CONDUCTION

When a solid is placed in an electrical circuit which maintains a voltage across it, the charged particles in the solid tend to rearrange themselves, that is, the anion and cation defects and the electrons in the solid move so as to let current flow in the external circuit.   In metals and semiconductors essentially all of the current is carried by electrons. However, in ionic solids at high temperatures the ions are more mobile than the tightly bound electrons.   Thus electricity is conducted through the solid by the diffusion of ions.

To derive an equation relating the conductivity and the diffusion coefficient, it is necessary to take account of the force exerted on the ions by the electric field.   Following the reasoning that led to the phenomenological equations of Sec. 4-4, we proceed as follows: In the absence of an electric field the condition for equilibrium is $\nabla \mu_j = 0$ for each component.   If we continue to use the definition $\mu_j = \mu_0 + kT \ln c_j$, the condition for equilibrium in the presence of an electric field $\nabla \phi$ is

$$\nabla \mu_j + q_j \nabla \phi = 0 \tag{5-4}$$

where q_j is the charge on the particle and $-q_j \nabla \phi$ is the force on it due to the field.   The flux is equal to the product of the number of particles per unit volume (c_j), their mobility (B_j), and the mean force on the particles.   Thus if we consider the case in which μ_j and ϕ vary only along the x axis

$$J_j = -B_j c_j \left(\frac{d\mu_j}{dx} + q_j \frac{d\phi}{dx} \right) \tag{5-5}$$

In the absence of a field, the flux of particles can also be expressed as

$$J_j = -D_j \frac{\partial c_j}{\partial x}$$

or, from Eq. (5-5),

$$J_j = -B_j c_j \frac{\partial \mu_j}{\partial x}$$

Now from the definition of μ_j

$$\frac{\partial \mu_j}{\partial x} = \frac{kT}{c_j} \frac{\partial c_j}{\partial x}$$

So these two equations for the flux will be equal if

$$B_j kT = D_j \tag{5-6}$$

This equation is referred to as the Einstein equation or the Nernst-Einstein equation.[1]

To demonstrate how D is related to the electrical conductivity σ, consider the case in which the charge is carried by interstitial atoms. The flux of charge, or the current I, will then be given by the equation $I = J_i q_i$, and if $(\partial c_i / \partial x) = 0$, the current will be

$$I = \frac{D_i q_i^2 c_i}{kT} \left(-\frac{d\phi}{dx} \right)$$

The conductivity is defined by the equation $I = \sigma(-d\phi/dx)$, so we can write

$$\frac{\sigma}{D_i} = \frac{c_i q_i^2}{kT} \tag{5-7}$$

However, using radioactive tracers, we do not measure the diffusion coefficient for the interstitials (D_i) but D_T, the diffusion coefficient for a radioactive tracer in the solid. These two can be related using the random-walk equations for D.† If an atom moves by jumping from one interstitial site directly to another, the random-walk equation gives

$$D_i = \tfrac{1}{6} \Gamma_i \alpha^2$$

---

[1] In the physics literature the term "mobility" is not applied to our B but to the product eB where e is the electronic charge. This product is designated by the symbol μ. This procedure is not followed here since the symbol μ is used to represent the chemical potential and since the mobility as defined in Eq. (5-6) has already been used in Chap. 4.

† The author is indebted to Dr. John Manning for helpful discussions of the following relation of σ to D_T.

where Γ_i is the jump frequency of an interstitial ion. Now a tracer atom moving in the same manner will spend only a small fraction of its time in interstitial positions. This time fraction is taken to be equal to c_i/c where c is the number of ions of the given type per unit volume.[1] Thus for diffusion by an interstitial mechanism

$$D_T = \frac{1}{6}\,\Gamma_i\frac{c_i}{c}\,\alpha^2$$

and

$$\frac{\sigma}{D_T} = \frac{cq_i^2}{kT} \quad \text{(interstitial)} \tag{5-8}$$

If a vacancy mechanism of diffusion is dominant instead of an interstitial one, Eq. (5-7) becomes

$$\frac{\sigma}{D_v} = \frac{c_v q_v^2}{kT} \tag{5-9}$$

From the random-walk analysis

$$D_v = \frac{1}{6}\,\Gamma_v\alpha^2$$

However, with a vacancy mechanism the jumps of a tracer atom are correlated and from Sec. 3-3

$$D_T = \frac{1}{6}f\Gamma_v\,\frac{c_v}{c}\,\alpha^2 = fD_v\,\frac{c_v}{c}$$

It follows that

$$\frac{\sigma}{D_T} = \frac{cq_v^2}{fkT} \quad \text{(vacancy)} \tag{5-10}$$

Thus we have the interesting result that the relation between σ and D_T varies with the mechanism of diffusion.[2] We will study this in detail in Sec. 5-6.

---

[1] If we follow one ion over a long period of time, it will spend a certain fraction of this period (t_i/t) in an interstitial position. The equivalence of this time average for one ion to the fraction of ions on interstitial sites at any instant (c_i/c) is best taken as being intuitively obvious. It is a simple case of the Ergodic theorem of statistical mechanics and is apparently correct though a rigorous proof is very difficult.

[2] An interesting paradox can be obtained by considering, instead of the vacancy flux, the equal and opposite flux of charged ions. This gives in place of Eq. (5-7)

$$\frac{\sigma}{D_a} = \frac{cq^2}{kT} \tag{5-11}$$

where D_a is the diffusion coefficient of the atoms (ions). Now it is a basic premise

Before the equation relating σ and D_T can be used, it is necessary to know the fraction of the observed conductivity that is due to the jth type of ion σ_j. The total conductivity σ is due to the movement of anions, cations, and electrons. The fraction of the total current carried by the jth type of particle is termed the transport number t_j. Thus

$$t_a + t_c + t_e = 1$$

where t_a is the fraction of the current carried by anions, etc. It follows that $\sigma_a = \sigma t_a$, etc. For alkali halide and silver halide crystals at temperatures above two-thirds of their melting point, t_e is negligible. It is also found that $t_c \simeq 1$ in these compounds, although t_a can be about unity in other compounds, for example, BaF_2, $BaCl_2$, PbF_2, $PbCl_2$.† This observation that either t_a or t_c is close to unity in most ionic crystals can be understood as follows. Both D_a and D_c will vary exponentially with temperature, and the activation energies for the two will be large but different. It follows that D_a can be much larger or much less than D_c over an appreciable temperature range, so in many cases the ionic conduction will occur through the movement of one type of ion.

5-3. EXPERIMENTAL CHECK OF RELATION BETWEEN σ AND D_T

The relation between σ and D_T indicated in Eq. (5-10) can be checked experimentally and has been for several cases. The results of a study on NaCl are shown in Fig. 5-2. Here the diffusion coefficient of the sodium ion (D_T) was determined by evaporating a thin film

---

of tracer experiments that radioactive isotopes and stable isotopes of the same atom diffuse in exactly the same manner, thus it would appear obvious that $D_T = D_a$. However, this result makes Eq. (5-11) disagree with Eq. (5-10), which requires that $D_a f = D_T$. In an attempt to understand how it is possible that D_a does not equal D_T, note that D_T and D_a measure two different quantities. D_T measures the rate at which two isotopes of a component intermingle in the absence of a net vacancy flux. D_a, on the other hand, measures how fast an atom moves to the left in a situation where vacancies are flowing from left to right. Thus it is not obvious that D_a must equal D_T, and since the reasoning leading from Eq. (5-9) to (5-10) seems to be unassailable, we must conclude that Eq. (5-10) is the correct one. (The reader who at this point feels too confused to trust to frail logic may take comfort in the fact that this conclusion will be buttressed by experimental facts in Sec. 5-6).

† A summary of transport number data as well as an extensive discussion of the entire field of ionic conduction is given by A. B. Lidiard in "Handbuch der Physik," vol. 20, p. 246, Springer-Verlag, Berlin, 1957.

of NaCl containing radioactive sodium on the surface of a single crystal. This was diffused, then sectioned, and D_T determined using the thin-film solution described in Sec. 1-3. In measuring the conductivity, the technique differs from that used on metals primarily in two respects. First, the currents measured are very small since σ is

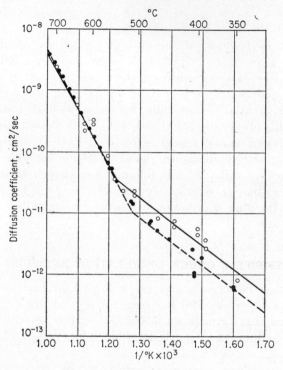

FIG. 5-2. Log D versus $1/T$ for sodium in NaCl as determined with radioactive sodium (o) and as calculated from the conductivity (●). [*From Mapother, Crooks, and Maurer, J. Chem. Phys.*, **18**: 1231 (1950).]

about 10^{-3} (ohm-cm)$^{-1}$ or 10^{-9} times that of copper at room temperature. Second, the voltage must be pulsed or alternated in sign so that the crystal will not become polarized at the electrodes.

In NaCl the sodium ions carry all of the current, and they move by a vacancy mechanism. Since the sodium ions of NaCl form an fcc sublattice, $f = 0.78$ and Eq. (5-10) predicts that the $D_{cond} = \sigma k T/ce^2$ should give a curve parallel to but 20% greater than the measured values of D_T. The results in Fig. 5-2 show that the agreement is good between the melting point and about 550°C, although the data are not accurate enough to say whether or not $D_{cond}/D_T = 1/f$. Below

550°C two changes occur: the slope changes appreciably, and the measured values of D_T are greater than those calculated from σ by a factor of 2. Both of these effects are due to the presence of metal impurities with a valence different from that of sodium. These effects will be explained in the next section.

5-4. EFFECT OF IMPURITIES ON D_T AND σ

The effect of impurities on the conductivity of ionic solids has been the subject of many studies and provides a powerful tool for studying the types and relative mobilities of the various defects formed. The power of the technique comes from the fact that there are definite relations between the impurity concentration and the defect concentration. For example, if a small amount, say 0.01%, of $CaCl_2$ is dissolved in NaCl, the solution can be thought to occur by either of two imaginary processes. First, two NaCl molecules could be removed per $CaCl_2$ molecule added. There would be no net change in the number of chlorine ions but, since the Ca^{++} ion occupies only one of the two vacated Na^+ sites, a cation vacancy will be introduced for each Ca^{++} ion added. A second possible process would be to remove only one NaCl molecule per $CaCl_2$ molecule and place the extra chlorine ion in an interstitial position. Of these two possible processes the latter would involve a larger free-energy charge than the former, that is, $\Delta G_{ia} > \Delta G_{vc}$. Thus, when $CaCl_2$ is added to NaCl, a cation vacancy will be added for each Ca^{++} ion. To maintain charge neutrality, the number of defects with a net positive charge (Ca^{++} ions and anion vacancies) must equal the number of defects with a net negative charge. The relation between the atom fraction of the impurities and the atom fraction of defects is therefore

$$N_{++} + N_{va} = N_{vc} \qquad (5\text{-}12)$$

where N_{++} is the fraction of cation sites occupied by divalent impurities.

The conductivity and diffusivity are both proportional to the concentrations of mobile defects so that the problem is to see how this concentration is varied by the impurity concentration. We shall discuss the case of Schottky disorder in the NaCl-type lattice, although the treatment of Frenkel disorder is analogous. In a pure NaCl-type crystal the concentration of anion vacancies ($N_{va}{}^0$) must equal that of the cation vacancies ($N_{vc}{}^0$). (The superscript zero is used to denote the pure material.) If divalent impurities are added, the concentration of cation vacancies and anion vacancies will be different, but

Eq. (5-2) is still valid. Thus using Eq. (5-12) to substitute for N_{va} gives

$$(N_{vc})(N_{vc} - N_{++}) = \exp\left(-\frac{\Delta G_S}{RT}\right) = N_{vc}{}^{02} = N_{va}{}^{02} \quad (5\text{-}13)$$

This equation assumes that the impurities and the defects are all randomly distributed, i.e., that the solution is ideal. (This is not strictly true, but it is a useful approximation.) Equation (5-13) can be rewritten

$$\left(\frac{N_{vc}}{N_{++}}\right)^2 - \frac{N_{vc}}{N_{++}} - \frac{N_{vc}{}^{02}}{N_{++}{}^2} = 0$$

This is a quadratic equation in (N_{vc}/N_{++}). Now N_{vc}/N_{++} must be positive, so the root of the equation is

$$N_{vc} = \frac{N_{++}}{2}\left[1 + \left(1 + \frac{4N_{vc}{}^{02}}{N_{++}{}^2}\right)^{\frac{1}{2}}\right] \quad (5\text{-}14)$$

This equation simplifies in two limiting cases—first, when $N_{vc}{}^0 \gg N_{++}$, then $N_{vc} = N_{vc}{}^0$; second, when $N_{vc}{}^0 \ll N_{++}$, then $N_{vc} = N_{++}$. The former case applies to pure material, while the latter case applies to impure material. However, $N_{vc}{}^0$ varies exponentially with temperature, and since no material is absolutely free from multivalent impurities, there will always be a temperature below which $N_{vc}{}^0 \ll N_{++}$.

We are now in a position to explain the change in slope at 550°C shown in Fig. 5-2. If sodium diffuses by a vacancy mechanism, D_T will be given by an equation of the form

$$D_T = \gamma a_0{}^2 N_{vc} w_T$$

where γ is a constant roughly equal to 1, and w_T is the jump frequency of a sodium tracer next to a vacant site. In the temperature range where $N_{++} \ll N_{vc}{}^0$, this equation can be rewritten

$$D_T = \gamma a_0{}^2 \exp\frac{-\Delta G_S}{2RT} \exp\frac{-\Delta G_m}{RT}$$
$$= D_0 \exp\frac{-\Delta H_S/2 - \Delta H_m}{RT} \quad (5\text{-}15)$$

The variation of D_T with temperature in this range stems from the fact that both N_{vc} and w_T vary with temperature. The observed D_T, and thus σ, are independent of the purity or history of the specimen in this range so that these properties are intrinsic properties of the compound; the name "intrinsic range" has been applied to the phenomenon since before it was understood.

In the low-temperature range $N_{++} \gg N_{vc}{}^0$. Here the vacancy concentration no longer varies with temperature but equals N_{++}. The equation for D_T is then

$$D_T = D_0' \exp\left(\frac{-\Delta H_m}{RT}\right) \tag{5-16}$$

where D_0' is much less than D_0 and is roughly equal to $N_{++} D_0$.

The validity of Eqs. (5-15) and (5-16) can be checked in at least two ways. First, the difference between the observed slope in Fig. 5-2 in the intrinsic range and that in the low-temperature or extrinsic range should equal $\Delta H_S/2$. The measured activation energies for the two ranges is shown in Table 5-1. The indicated value of ΔH_S is 2.06 ev. This agrees with the calculated value of 1.92 ev.[1] The values of the ratios of D_0 to D_0' are also given. The crystals used should be better than 99.99% pure; that is in agreement with the observed values of D_0/D_0' and the prediction that $D_0/D_0' \simeq 1/N_{++}$.

TABLE 5-1. Diffusion Data for Sodium in NaCl

Intrinsic range:

D_0 (cm^2/sec)	3.1[a]
$H_m + H_S/2$ (ev)	1.80[a]

Extrinsic range:

D_0' (cm^2/sec)	1.6×10^{-6a}
H_m (ev)	0.77[a]
H_S (ev)	2.06
D_0/D_0'	2×10^6

[a] D. Mapother, H. Crooks, and R. Mauer, *J. Chem. Phys.*, **18**: 1231 (1950).

The other check on this model comes from measuring the conductivity of a series of crystals which contain controlled amounts of divalent impurities. If the model leading to Eq. (5-16) is correct, the values of σ, or D_T, should increase linearly with N_{++} at any given temperature in the extrinsic range. Experiment shows that this is essentially the case. Values of σ for NaCl doped with CdCl$_2$ are shown in Fig. 5-3. The values of D_0' increase with N_{++}, while the slope is unchanged. Also when the temperature is high enough so that the extrinsic conductivity of each crystal becomes less than that of pure NaCl, the slope increases, and the intrinsic conductivity becomes dominant.

[1] M. Tosi and F. Fumi, quoted by Lidiard, *op. cit.*, p. 274.

If σ of the above study is plotted against N_{++}, that is, N_{Cd}, at some fixed temperature in the extrinsic range, it is found that the line is not quite straight but is concave downward. Thus σ does not quite increase linearly as expected. The number of vacancies must still be equal to N_{++} so all of the vacancies must not be contributing to σ. This ineffectiveness of some vacancies stems from the fact that there is

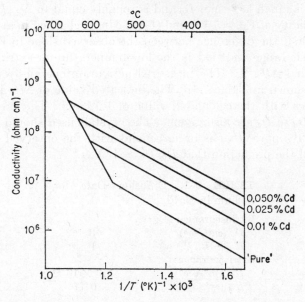

Fig. 5-3. Log σ versus $1/T$ for NaCl crystals doped with varying atom per cent of CdCl$_2$.

a coulombic attraction between the divalent impurities and the vacancies. Thus there will be an equilibrium of the form

$$Cd^{++} + \square \rightleftharpoons (Cd^{++}\square)$$

where $(Cd^{++}\square)$ designates a cation vacancy on a site next to a cadmium ion and loosely bound to it.

It is customary to think of the cadmium ion in NaCl as having a unit of positive charge associated with it and of the cation vacancy as having a unit of negative charge associated with it. This is what gives rise to the attraction between the two. The two together will have no net charge, and the bound vacancy will not migrate toward the anode as an unassociated vacancy would. Since it is the migration of cation vacancies toward the anode which gives rise to the conductivity, it follows that the vacancies which are associated with, or "bound" to, divalent impurities do not contribute to σ as do the unassociated ones.

This association of vacancies and divalent impurities also explains the observation (Fig. 5-2) that in the extrinsic range D_T is appreciably larger than the value calculated from σ. Though a bound vacancy will not migrate in a field, and thus will not contribute to σ, it will move around through the crystal and contribute to the intermixing of radioactive and stable sodium ions. The rate of this intermixing is indicated by D_T, so the bound vacancies as well as the free vacancies contribute to D_T.[†] In the intrinsic range $N_{vc}^0 \gg N_{++}$, so that the number of bound vacancies will not be large enough to influence σ.

5-5. EFFECT OF IMPURITIES ON CONDUCTIVITY IN CRYSTALS WITH FRENKEL DISORDER

In the example of Schottky disorder studied above, the mobility of the cation vacancies was so much greater than that of the anion vacancies that the movement of the latter defect could be ignored. In the best-studied materials which exhibit Frenkel disorder, the conductivity occurs entirely by cation movement, that is, $t_c = 1$. However, the cations move both by a vacancy and by an interstitial mechanism. We shall see in the next section that the unambiguous relation of D_T to σ requires a knowledge of the relative mobilities for these two defects. Thus we shall give a qualitative discussion of how the relative mobilities can be determined, before deriving the equations relating σ and D_T.

To a good approximation the interstitials and vacancies move independently so that the total conductivity due to the cations can be taken as the sum of the conductivities of each type of defect; thus

$$\sigma = \sigma_i + \sigma_v$$
$$\sigma = c_i q_i^2 B_i + c_v q_v^2 B_v \tag{5-17}$$
$$\sigma = c_c q_c^2 (B_i N_i + B_v N_v) \tag{5-18}$$

where c_c is the number of cations per unit volume, and we have used the relations $c_i = c_c N_i$, $c_v = c_c N_v$ and $q_c = q_i = -q_v$.[‡]

If the ratio N_i/N_v is varied without changing the ratio B_i/B_v, the change of σ with N_i/N_v will indicate whether or not the defect being added is the more mobile. This can be done as follows: if small

[†] For a calculation of σ/D_T in this case, see Lidiard, *op. cit.*, p. 332.

[‡] The equations developed here are for the case of a compound made up of monovalent ions, for example AgBr.

amounts of $CdBr_2$ are added to AgBr, the various concentrations (atom fractions) must obey the equations

$$N_{++} + N_{ic} = N_{vc}$$

$$(N_{vc})(N_{ic}) = N_{vc}{}^{02} = N_{ic}{}^{02} = \exp\left(-\frac{\Delta G_F}{RT}\right) \qquad (5\text{-}3)$$

Thus as the concentration of divalent cadmium ions is increased, the concentration of interstitials will decrease, and that of the vacancies will increase. Experimentally, it is found that the conductivity decreases when $CdBr_2$ is first added to pure AgBr. The conclusion is that the silver interstitials are more mobile than the silver vacancies.

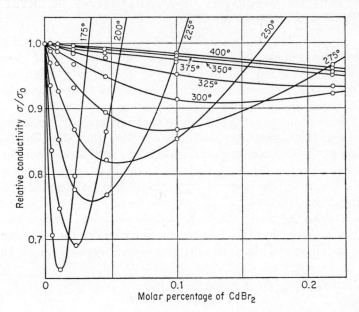

Fig. 5-4. Relative conductivity (σ/σ_0) of AgBr vs. atom fraction of $CdBr_2$ at several temperatures. [*After J. Teltow, Ann. Physik,* **5**: 63 (1949).]

If the concentration of $CdBr_2$ is increased, σ continues to decrease as long as the majority of the conduction is due to interstitials. However, as N_{++} and N_{vc} continue to increase, there comes a composition at which σ no longer decreases but goes through a minimum and starts to increase (see Fig. 5-4). In this range the vacancies are starting to give the majority of the conduction. As N_{++} is increased further, σ increases. Using the concepts developed above, an equation can be obtained which expresses σ/σ^0 in terms of N_{++}, $N_{vc}{}^0$, and B_i/B_v.[†] (σ^0 is σ at $N_{++} = 0$.) Measurements of σ/σ_0 versus N_{++} at various temperatures then allow the determination of B_i/B_v and $N_{vc}{}^0$ from the values of σ/σ_0 and N_{++} at the minimum.   The equations are

$$(N_{++})_{\min} = \frac{N_{vc}{}^0(\phi - 1)}{\sqrt{\phi}} \qquad (5\text{-}19)$$

$$(\sigma/\sigma_0)_{\min} = \frac{2\sqrt{\phi}}{1 + \phi} \qquad (5\text{-}20)$$

where $\phi = B_i/B_v$. Figure 5-4 shows that $(\sigma/\sigma_0)_{\min}$ increases as the temperature increases.   In view of Eq. (5-20), this means that ϕ decreases with increasing temperature.   Teltow found that for AgBr, ϕ changes from 7 at 180°C to 2 at 350°C.   Since the temperature dependence of each defect mobility is determined entirely by the activation energy for a defect jump, the temperature dependence of ϕ comes from the difference in ΔH_m for the two defects.   From the two data above, it can be calculated that $\Delta H_m(v) \simeq \Delta H_m(i) + 0.2$ ev.

5-6. RELATION OF σ TO D_T IN AgBr (FRENKEL DISORDER)

Careful measurements on σ and D_T for silver in AgBr[‡] have shown that D_{cond}, that is, $\sigma kT/ce^2$, ranges from over twice D_T at low temperatures to $1.5D_T$ at high temperatures.   This ratio is even larger than the deviation predicted for the vacancy mechanism [Eq. (5-10)]. Thus a mechanism other than interstitial or vacancy is implied.   Of the mechanisms we have studied, the ring or pair-exchange mechanism is ruled out since if it were dominant, σ would be zero while D_T would be finite, and D_{cond}/D_T would equal zero.   Another mechanism which has been suggested is the interstitialcy mechanism.   We shall show below that it can give up to $D_{\text{cond}}/D_T = 3$, thus making it the most probable mechanism for the movement of the interstitial silver ions.

An interstitial ion can move by jumping directly to another interstitial site (interstitial mechanism) or by pushing one of its nearest-neighbor ions into an interstitial position and taking its place (interstitialcy mechanism).   If the latter occurs, the interstitial in effect moves farther than either of the two ions involved.   Figure 5-5 shows an interstitial ion in one of the eight interstitial positions in an AgBr

† J. Teltow, *Ann. Physik. Leipzig*, **5**: 63, 71 (1949).   This is in German.   For its discussion in English, see Lidiard, *op. cit.*, p. 288.

‡ R. Friauf, *Phys. Rev.*, **105**: 843 (1957).   This has also been shown for AgCl by W. Compton, *Phys. Rev.*, **101**: 1209 (1956).

unit cell.   A heavy arrow indicates a jump to its nearest neighbor in
the exact center of the cell.   This center ion can be displaced in a
forward direction to the center of any one of four cubes.   If it goes in
the direction of the heavy arrow, [111], the jump will be called a
collinear jump.   If it goes to one of the other three forward positions,
it will be called a noncollinear jump.

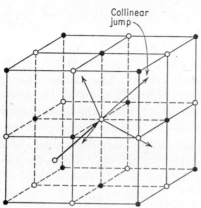

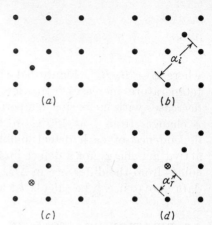

FIG. 5-5. The heavy arrow shows the
movement of the interstitial atom
making a jump by the interstitialcy
mechanism.   The other four arrows
represent the possible jumps of the
atom displaced from a normal lattice
position.   The collinear jump vector
is so labeled; the three possible noncollinear jump vectors are not.

FIG. 5-6. (a) and (b) show schematic
diagrams of a two-dimensional square
lattice before and after a collinear
interstitial jump has occurred. α_i
is the distance which the interstitial
moves in this process.   (c) and (d)
show the same process, except that
a tracer atom ($\otimes$) is involved. α_T is
the distance which the tracer moves
in this process.

To relate D_T to σ for the interstitialcy mechanism, we again start
with Eq. (5-8)

$$\frac{\sigma}{D_i} = \frac{c_i q_i{}^2}{kT} \tag{5-7}$$

where D_i is the diffusion coefficient for the interstitial ion.   The cation
sites neighboring an interstitial cation are all equivalent, so that all
jump directions are equally probable and successive jump directions of
the interstitial are uncorrelated.   Thus

$$D_i = \tfrac{1}{6}\Gamma_i \alpha_i{}^2$$

It should be emphasized that this equation refers to the diffusion of the
interstitial.   To clarify the relation between the movement of the

interstitial and the movement of a particular ion, consider the two-dimensional square lattice shown in Fig. 5-6. Between Fig. 5-6a and b the interstitial has moved by a collinear interstitialcy mechanism. As far as a conductivity experiment can indicate, the only change between (a) and (b) is that the charged interstitial has moved a distance α_i. The same process is shown again in Fig. 5-6c and d, except that here a tracer is involved. In a tracer experiment a particular ion is followed instead of the interstitial, and the distance a tracer ion moves in the jump (α_T) is one-half of α_i.

The equation for D_T is

$$D_T = \tfrac{1}{6} f \Gamma_T \alpha_T{}^2$$

That $f < 1$ in this case can be seen from Fig. 5-6c and d. In c the neighbors of the tracer are identical, so there can be no correlation between the tracer's last jump and its next. In Fig. 5-6d the tracer must make its next jump to an interstitial site, and these are not all identical. Thus in (d) it is more probable that the tracer will make its next jump backward in the direction it just came from than forward in the direction of its last jump. This correlation between jumps makes $f < 1$.

To derive an equation relating D_i to D_T for the collinear interstitialcy mechanism, we note first that $2\alpha_T = \alpha_i$. The relation between Γ_T and Γ_i can be obtained as follows: If over a long period of time t a tracer ion makes n jumps, n_i of which are from interstitial sites and n_n from normal sites, then

$$\Gamma_T = \frac{n}{t} = \frac{n_i}{t} + \frac{n_n}{t}$$

But $n_i = n_n$, since this mechanism requires that the tracer always jump from a normal site to an interstitial, or vice versa. If t_i is the time spent on interstitial sites, this gives

$$\Gamma_T = \frac{2n_i}{t} = \frac{2n_i}{t_i} \frac{t_i}{t}$$

Now $n_i/t_i = \Gamma_i$ by definition and taking $t_i/t = c_i/c$ (see footnote 1, page 142) gives

$$\Gamma_T = 2\Gamma_i c_i/c \qquad (5\text{-}21)$$

For the collinear interstitialcy mechanism then

$$\frac{D_i}{D_T} = \frac{2c}{f c_i}$$

Detailed calculation shows that in this case $f = \frac{2}{3}$. Thus

$$\frac{\sigma}{D_T} = 3\frac{cq_i^2}{kT} \qquad \text{(collinear int.)} \qquad (5\text{-}22)$$

or $\qquad \sigma kT/cq_i^2 \equiv D_{\text{cond}} = 3D_T \qquad \text{(collinear int.)} \qquad (5\text{-}23)$

For a noncollinear mechanism, again $\Gamma_i c_i/c = 2\Gamma_T$, but $(\alpha_i/\alpha_T)^2 = \frac{8}{3}$ instead of 4. The effect of correlation is less in this case, and the result is

$$D_{\text{cond}} = 1.38 D_T \qquad \text{(noncollinear int.)} \qquad (5\text{-}24)$$

The third mechanism that contributes to diffusion in AgBr is a vacancy mechanism. This was discussed above. Since $\alpha_a = \alpha_v$, the

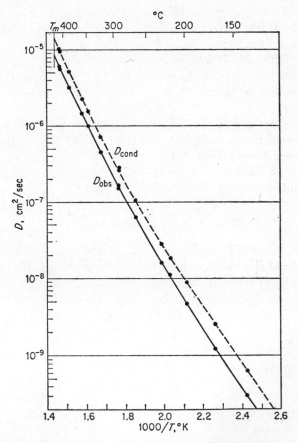

Fig. 5-7. Log D versus $1/T$ for silver in AgBr as determined by tracer and conductivity experiments. [*From R. J. Friauf, Phys. Rev.*, **105**: 843 (1957).]

entire effect is due to correlation, and from Eq. (5-10), we have

$$D_{\text{cond}} = 1.27 D_T \quad \text{(vacancy)} \quad (5\text{-}24a)$$

Experimental results for AgBr are shown in Fig. 5-7. The ratio D_{cond}/D_T varies from 2.17 (140°C) to 1.50 (350°C). It is apparent from the large value of D_{cond}/D_T that the collinear interstitialcy mechanism must account for an appreciable part of the conductivity at 140°C, and for a smaller, but finite, fraction at the higher temperature. With the help of Teltow's results on the relative mobilities of vacancies and interstitials in AgBr (see Sec. 5-5), Friauf actually was able to determine the relative contributions of collinear and noncollinear jumps.

In the last two sections, we have shown that the relation between σ and D_T and the requirement of charge neutrality combine to give a unique set of tools for studying diffusion in ionic materials. Charge neutrality allows the concentration of defects to be varied in a known manner, while the relation of σ to D_T allows the unambiguous determination of the mechanism of diffusion. The situation in metals is not nearly so satisfactory. There one can infer some things from marker movement studies, but in the final analysis it is necessary to fall back on the calculation of energy terms from theoretical models. These calculations give unambiguous results for copper and silver, but in going to less noble metals the evidence dwindles from masses of circumstantial evidence to mere guesses.

5-7. DIFFUSION IN SEMICONDUCTORS

Though the term "semiconductor" refers to the mechanism of electrical conductivity in a solid, solids which show this characteristic also have similar diffusion characteristics. In semiconductors the concentration of mobile, or conducting, electrons is high enough to make the electronic transport number equal to unity but low enough that they can be treated as independent particles. Thus the simple relation between σ and D_T found in ionic conductors is lost, but the concept of charge neutrality is still helpful.

Since we are interested only in the properties unique to nonmetals, the discussion will deal primarily with compounds. One of the characteristics of many semiconducting compounds formed by transition metals is an appreciable deviation from stoichiometry. A well-studied and extreme example of this is provided by the wustite or FeO phase of the iron-oxygen system. At 1200°C this phase exists over a composition range of roughly 51 to 54 atom per cent oxygen.[1] Thus the compositions deviate markedly from the stoichiometric FeO composition. This deviation is obtained by creating vacancies on the iron ion sites, the composition range at 1200°C corresponding to 2 to 8% of the iron sites being vacant. Charge neutrality is maintained by creating two Fe^{3+} ions for each Fe^{++} vacancy.[2]

In wustite (FeO) the iron ion diffuses by a vacancy mechanism. Thus the diffusion coefficient of an iron tracer will be given by an equation of the form

$$D_T = \gamma a_0^2 N_v w_T \qquad (5\text{-}25)$$

where γ is a constant and N_v is the fraction of vacant cation sites. Because of the large deviation from stoichiometry in wustite, N_v ranges only between 0.02 and 0.08 and for a given composition is independent of the temperature. The variation of D_T with temperature at any given composition will thus stem from ΔH_m in the equation for the jump frequency. This is similar to the behavior of D, or σ, in the extrinsic region of an ionic material, except that in wustite the vacancies are caused by chemical forces instead of divalent impurities.

Diffusion data for iron in various compositions of wustite are shown in Fig. 5-8. An equation for the slope of these lines can be obtained by differentiating Eq. (5-25) to give

$$\left(\frac{\partial D_T}{\partial N_v}\right)_T = \gamma w_T a_0^2 \qquad (5\text{-}26)$$

Thus the slope of each line is proportional to w_T, the jump frequency of an iron ion into a neighboring vacant site. The activation energy for w_T, that is, ΔH_m, could be obtained from the variation of the slope of these lines with temperature. However, a more accurate value is obtained by choosing a composition and taking the value of D_T for that composition from each line. A plot of log D_T versus $1/T$ then gives

$$\left(\frac{\partial \ln D_T}{\partial 1/T}\right)_{N_v} = \frac{d \ln w_T}{d 1/T} = -\frac{\Delta H_m}{R}$$

[1] For a discussion of the iron-oxygen phase diagram, see L. Darken and R. Gurry, "Physical Chemistry of Metals," pp. 350–359, McGraw-Hill Book Company, Inc., New York, 1953.

[2] The ease of ionizing Fe^{++} ions to Fe^{3+} is directly related to the occurrence of nonstoichiometry. If a compound is not stoichiometric, the excess or deficit of charge must be compensated for by changing the valence of the ions involved. In FeO this is easily done; in AgBr the energy required is quite large. Thus the appreciable deviation in FeO and lack of it in AgBr.

The first equality follows from the differentiation of Eq. (5-25); the second from the definition, $w_T = \nu \exp(-\Delta G_m/RT)$. For iron in FeO at Fe/O = 0.907, the equation obtained is[1]

$$D_{Fe} = 0.12 \exp \frac{-29,700}{RT} \text{ cm}^2/\text{sec} \qquad (5\text{-}27)$$

Data such as the above yield ΔH_m very simply, but the number of oxides with suitably large phase fields is rare.

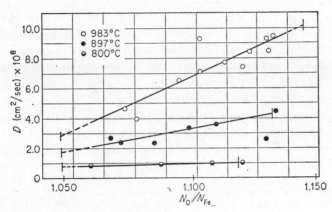

FIG. 5-8. Variation of the self-diffusion coefficient of Fe in FeO with composition. [*From L. Himmel, R. F. Mehl, and C. E. Birchenall, Trans. AIME*, **197**: 827 (1953).]

One of the most important characteristics of diffusion in compound semiconductors is the variation of D_T with the atmosphere. D_T is directly proportional to the defect concentration, and the defect concentration (composition) of the compound can be changed through the composition of the atmosphere. If the compound has close to the stoichiometric composition, a striking change can be made in the defect concentration with only a small change in composition. For example, if the composition of the compound MO is varied from M/O = $1 - 10^{-5}$ to M/O = $1 - 10^{-4}$, the change in composition is barely perceptible, but D_T could change by an order of magnitude. This type of behavior has been studied primarily in connection with oxidation phenomena and was pioneered by Wagner.[2]

[1] L. Himmel, E. Birchenall, and R. F. Mehl, *Trans. AIME*, **197**: 827 (1953).

[2] A complete review of this is given by K. Hauffe, "Reaktionen in und an festen Stoffen," Springer-Verlag, Berlin, 1955. Also see O. Kubaschewski and B. Hopkins, "Oxidation of Metals and Alloys," Academic Press, Inc., New York, 1953.

As an example of this type of behavior and the information obtainable therefrom, consider the case of CoO. The variation of vacancy concentration with oxygen partial pressure can be obtained by an equation of the form

$$2Co^{++} + \tfrac{1}{2}O_2 \rightleftharpoons 2Co^{3+} + V_{Co} + O^{--} \tag{5-28}$$

where V_{Co} designates a cobalt ion vacancy The electron deficit which converts a Co^{++} ion into a Co^{3+} ion is called an electron hole ($\oplus$). This is essentially an electron vacancy. At any instant it is associated with one ion, but it can move by an electron jumping from an adjacent Co^{++} ion to the Co^{3+} ion. The movement of electrons by the movement of electron holes is thus analogous to the diffusion of atoms by a vacancy mechanism. Similarly the electron hole can be treated as a distinct defect, and one can rewrite the above equilibrium in the form

$$\tfrac{1}{2}O_2 \rightleftharpoons 2\oplus + V_{Co} + O^{--} \tag{5-29}$$

If it is assumed that none of these species interact, i.e., the solution is ideal, the equilibrium constant for the reaction is[1]

$$\frac{N_{\oplus}{}^2 N_v}{P_{O_2}{}^{\frac{1}{2}}} = \beta^3 \exp\left(-\frac{\Delta H^0}{RT}\right) \tag{5-30}$$

where β is a constant and ΔH^0 is the molar enthalpy change for the reaction. Since a Co^{++} ion is removed to form each cation vacancy, charge neutrality requires that $N_v = 2N_{\oplus}$. Inserting this relation in Eq. (5-30) leads to

$$N_v = P_{O_2}{}^{\frac{1}{6}}\beta' \exp\left(-\frac{\Delta H^0}{3RT}\right) \tag{5-31}$$

If this equation is substituted in the expression $D_T = \gamma a_0{}^2 N_v w_T$, two results follow. First, the diffusion coefficient at constant temperature should increase linearly with $P_{O_2}{}^{\frac{1}{6}}$. Second, the quantity ΔH in the expression

$$\left(\frac{\partial \ln D_T}{\partial 1/T}\right)_{P_{O_2}} = -\frac{\Delta H}{R}$$

will equal $\Delta H_m + \Delta H^0/3$, instead of ΔH_m. This reflects the fact that at constant oxygen pressure the composition of the oxide changes with temperature.

[1] Since the concentration of O^{--} ions is not changed appreciably in this reaction, their concentration does not enter the equation. (In terms of chemical thermodynamics, the activity of the O^{--} ions is taken as unity.)

Experimental results for D_{Co} in CoO are shown in Fig. 5-9. The diffusion coefficient does increase with pressure, but it does not increase at the $\frac{1}{6}$th power of P_{O_2}. Instead it was found that $D = (\text{const})\,P_{O_2}^{1/n}$ where n equals 3.1 at 1000°C, 3.3 at 1200°C, and 3.6 at 1350°C.

The reason for this deviation of n from 6 is not well understood, although it is thought to be due to an interaction between the electron

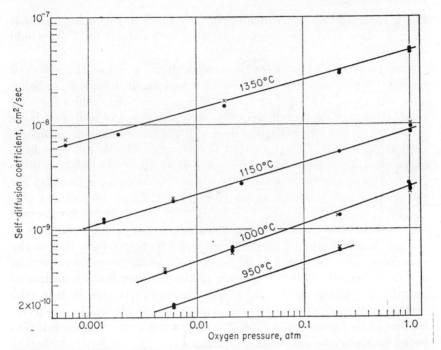

FIG. 5-9. Self-diffusion coefficient of Co in CoO as a function of partial pressure of oxygen. [*From R. Carter and F. Richardson, Trans. AIME*, **200**: 1244 (1954).]

holes and the vacancies. There can be an appreciable coulombic attraction between the positively charged electron holes and the vacancies (which have an effective charge of -2 due to the uncompensated oxygen ions around each). In view of this attraction, the holes and vacancies will not be randomly mixed, but there will be a high probability of a vacancy having at least one electron hole on one of its nearest-neighbor sites. If this situation is approximated by the case in which half of the electron holes are randomly distributed and the other half are bound to vacancies to form vacancy-hole complexes $(V_c \oplus)$, Eq. (5-29) becomes

$$\tfrac{1}{2}O_2 \rightleftharpoons \oplus + (V_c \oplus) + O^{--} \tag{5-32}$$

But now charge neutrality requires that $N_\oplus = N_{v\oplus}$, so

$$\frac{N_\oplus N_{v\oplus}}{P_{o_2}^{\frac{1}{2}}} = K(T)$$

It follows that for this case

$$N_{v\oplus} \approx P_{o_2}^{\frac{1}{4}}$$

If it is assumed that all electron holes are bound to vacancies, a similar analysis shows that the concentration of such defects ($\oplus V \oplus$) would vary as $P_{o_2}^{\frac{1}{2}}$.

This provides the basis for a plausible explanation of the observed results. If the association of a vacancy and an electron hole, or holes, does not affect the movement of the vacancies, the results for CoO would indicate that each vacancy has between one and two holes associated with it, and that the degree of association decreases slightly with temperature. To show that the associated holes will have little effect on the vacancy movement, we must know the relative jump frequencies of the two defects. It is possible to define a mobility for the electron hole in the same manner as one talks of the mobility of a vacancy. Since the electric current in CoO is all carried by electron holes whose concentration is comparable to that of the vacancies, it follows that the mobility of the holes is much greater than that of the vacancies. But the mobility is proportional to the jump frequency, so it follows that the jump frequency of the electron holes is orders of magnitude greater than that of the vacancy. If this is true, the complex will move at a rate which is determined only by the jump frequency of the vacancy. (The diffusion of the hole-vacancy complex would be similar to that of the impurity-vacancy complexes studied in Chap. 3.)

The effect of atmosphere on D_T has been studied primarily in connection with the theory of the oxidation of metals. Since the rate of oxidation is determined by the rate of movement of ions through the oxide, these studies bear directly on the variation of oxidation rate with the atmosphere. The existence of the effect is well catalogued, but detailed studies of the association of defects and their effect on D_T are rare.[1]

Elemental Semiconductors. The majority of the research work concerned with the transistor has been done on germanium and silicon. The electrical conductivity in these elements increases exponentially with temperature, it has an intrinsic range at high temperatures, and

[1] A short review of the diffusion data in oxides is given by C. E. Birchenall, *Met. Revs.*, **3**: 325 (1958).

an extrinsic, or impurity-controlled, range at low temperatures. In the extrinsic range the mobile electrons come from the ionization of impurities with valence greater than 4 (called donors) or the movement of electron holes introduced by impurities of valence less than 4 (called acceptors). The manufacturing of transistors entails the development of the proper distribution of added impurities. In many cases this distribution is obtained by diffusion. Much of the work that has been done on diffusion in these elements has thus been aimed at obtaining values of D for the impurities of interest. This involves no concepts that we have not covered before, and the results and procedures can be obtained with the help of the available review articles.[1]

The novel effects which are peculiar to these materials occur because of the effect of one solute on the diffusion of another. Thus they are best classified as ternary effects although they appear at such low-solute concentrations that the "ternary alloy" has a lower total impurity content than most samples used for pure metal self-diffusion studies. They will only be mentioned here along with references to articles giving a detailed discussion.

The first of these ternary effects is due to the interaction between ionized acceptors and donors. Consider an interstitial impurity which is attracted to a substitutional impurity. The effective jump frequency of the interstitial atom will be appreciably reduced when it is near the substitutional impurity. Thus the diffusion coefficient of the interstitial will be markedly influenced by the concentration of the substitutional impurity.

A second effect appears if an impurity of valence other than 4 is diffused into very pure material. The solute ionizes at the high diffusion temperature, and the electrons, or holes, become the third component. The electrons or holes are much more mobile than the solute ions, so they diffuse into the neutral solvent faster than the ions. However, this separates positive and negative charge and sets up a local field. The result is that the ions diffuse inward under the effect of a concentration gradient, plus the electrostatic attraction of the electrons, or holes, which have diffused farther into the material.[2]

---

[1] For results in germanium and silicon, see chapter by H. Reiss and C. Fuller in N. B. Hannay (ed.), "Semiconductors," Reinhold Publishing Corporation, New York, 1959. A review of the diffusion studies in compounds of interest for use in semiconducting devices is found in P. H. Sutter's chapter in R. Heikes and R. Ure (eds.), "Thermoelectricity: Science and Engineering," p. 154, Interscience Publishers, Inc., New York, 1961.

[2] Both of these effects are discussed by Reiss, Fuller, and Morin, *Bell System Tech. J.*, **35**: 535 (1956) (Sec. 11 and Appendix), and by S. Zaromb, *IBM J. Res. Dev.*, **1**: 57 (1957).

Another phenomenon, which has been studied only in germanium, occurs in the case of copper or nickel diffusing in germanium. It appears that these impurities can go into solution either interstitially or substitutionally. The interstitial solubility N_i is much lower than the substitutional solubility N_s. However, the mobilities B_i and B_s of the two defects are such that $B_iN_i \gg B_sN_s$. Thus the solute diffuses into the germanium rapidly as an interstitial ion and then reacts with a vacancy to go to a substitutional position. If there are many active sources of vacancies, no anomalous behavior occurs. However, in the high-perfection germanium crystals available, there are often not enough dislocations to provide the required number of vacancies. As a result, the observed diffusion coefficient varies markedly with the perfection of the crystal; this can lead (and has) to conflicting results.[1] This variation of D with crystal perfection also leads to a violation of Fick's first law, since the ratio of the flux to the concentration gradient, that is, D, is not a constant or simply a function of composition.

5-8. DIFFUSION IN ORDERED ALLOYS AND INTERMETALLIC COMPOUNDS

Ordered alloys and intermetallic compounds form the bridge between dilute alloys in which the tendency toward short-range order is weak and the nonmetallic compounds in which one type of ion diffuses only on its own sublattice. In an ordered alloy the energy of an A-type atom on a B-type site is higher than on an A-type site. However, this energy difference is small enough that some A atoms do reside on B-type sites. (This can be seen from the lack of perfect long-range order or from the existence of the ordered alloy over a range of compositions.) The study of diffusion in such a lattice involves two difficult problems. First, the vacancy concentration will depend on the composition of the phase, but one must determine to what extent a deviation from stoichiometry toward an excess of A is accommodated by placing A atoms on B sites and to what extent by forming vacant B sites. Second, if A atoms diffuse by moving into vacant adjacent B-type sites, there will be a high probability of the A atom returning to the vacant A site on its next jump. This correlation effect will be very important. In fact, Flinn and McManus have concluded that in ordered β-brass the correlation effect is so strong that the *effective*

[1] A review of the literature on this subject is given by H. Van Bueren, "Imperfections in Crystals," 2d ed., chap. 31, North-Holland Publishing Company, Amsterdam, 1961.

diffusion jumps occur only by diffusion to second-nearest-neighbor positions; i.e., the Zn atoms diffuse primarily on their own sublattice.[1] Such second-nearest-neighbor jumps would have a larger activation energy than the nearest-neighbor jumps. And an experimental study of self-diffusion in β-brass shows that the activation energy increases in going from the disordered to the ordered phase.[2]

There has been little systematic diffusion work done on this class of compounds, so solutions to these problems have yet to be established.

PROBLEMS

5-1. A diffusing ion's free energy is decreased by $\alpha e \dfrac{d\phi}{dx} = e\Delta\phi$ in moving from left to right in Fig. 5-10. If the free energy of activation for the jump is Δg and the vibration frequency is ν:

(a) Show that the jump frequency from left to right is $\nu \exp\left[-\Delta g + e(\Delta\phi/2)\right]/kT$ and from right to left is $\nu \exp\left[-\Delta g - e(\Delta\phi/2)\right]/kT$.

(b) The conditions for a normal ionic conductivity experiment might involve $d\phi/dx = 10$ volts/cm and $T = 1000°$K. If $\alpha = 2 \times 10^{-8}$ cm, $\Delta g = 2$ ev, and $\nu = 10^{13}$ sec^{-1}, calculate the mean jump frequency and the ratio of the two jump frequencies.

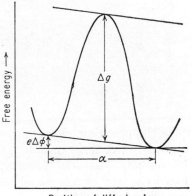

Position of diffusing ion

Fig. 5-10

5-2. In the high-temperature homogenization of a pressed mixture of AgCl and AgBr powders, what is the relation between the electrical conductivity and the rate of homogenization? (The AgCl and AgBr will form a single-phase mixture in the proportions used here.)

5-3. Find the ratio of N_{vc}/N_{va} under the conditions:

(a) That the free energy to form Frenkel defect pairs (ΔG_F) equals that required to form Schottky defect pairs (ΔG_S).

(b) That $\Delta G_F - \Delta G_S = 10$ kcal/mole and $T = 600°$K.

(c) That $\Delta G_F - \Delta G_S = 10$ kcal/mole and $T = 1000°$K.

5-4. There is a coulombic attraction between divalent impurities and vacancies in NaCl. If the jump frequency of a vacancy on any cation site is $12\ w_v$ and that of a Ca^{++} ion next to a vacancy is $12\ w_{++}$, explain what is wrong, if anything, with taking

$$D_{++} \simeq a_0{}^2 w_{++} p_v$$
$$D_v \simeq a_0{}^2 w_v$$

(a) In the intrinsic range.

(b) In the extrinsic range.

[1] P. Flinn and G. McManus, *Phys. Rev.*, **124**: 54 (1961).
[2] A. B. Kuper et al., *Phys. Rev.*, **104**: 1536 (1956).

chapter 6 HIGH-DIFFUSIVITY

PATHS

In the preceding chapters we have discussed only diffusion through crystals which were considered to be perfect aside from vacancies or interstitials. In any real crystalline specimen there are also dislocations, free surfaces, and often grain boundaries. It is now well established that the mean jump frequency of an atom in these regions is much higher than that of an atom in the lattice. The diffusivity is therefore higher in these regions. This higher diffusivity will be of interest for two reasons. First, there is the question of how much these paths contribute to the measured values of the diffusion coefficient. We assumed that they make no contribution, but we can lay down the conditions under which this will be true only after we know more about the effect of grain boundaries and dislocations. The second reason for interest in these high-diffusivity paths is that, in properly designed experiments, it is possible to determine the diffusion coefficient in each of these regions. By measuring this D in various types of grain boundaries, surfaces, or dislocations, it is possible to learn more about the structure of these paths and about how the atoms move in them.

As an example of the phenomena we are talking about, the contribution of diffusion along grain boundaries can be seen in Fig. 6-1. Here the apparent self-diffusion coefficient in silver is shown for single-crystal and polycrystal samples. This apparent diffusion coefficient is just that value of D obtained by plating radioactive silver on the surface of the specimen, diffusing it, and then determining D from a plot of ln activity vs. penetration distance squared. At high temperatures, the same value of D is obtained for both types of samples. However, below 700°C the values of D obtained using a polycrystal specimen consistently lie above the values obtained with a single crystal. The high-diffusivity paths[1] in this case are grain boundaries. It is seen that the contribution of the grain boundary regions is measurable around 600°C and becomes dominant below this temperature.

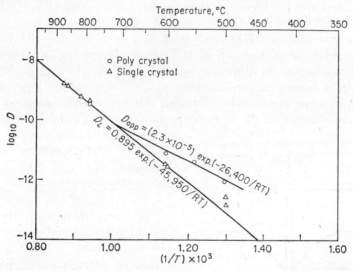

FIG. 6-1. Values of the self-diffusion coefficient obtained for silver using single-crystal and polycrystal samples. (*After D. Turnbull, in "Atom Movement,"* p. 129, *ASM, Cleveland,* 1951.)

An estimate of the increase in the jump frequency in the neighborhood of a grain boundary can be obtained as follows. In pure silver the smallest grain diameter which can be retained at high temperatures will be about 1 mm. If this is true and the high-diffusivity region around a grain boundary is taken to be 3×10^{-8} cm wide, about one atom in 10^6 will be in the grain boundary. At 650°C these few grain

[1] Zener coined the phrase "short-circuiting paths" to describe this type of effect, and it is commonly referred to in this manner. If the reader is accustomed to think in terms of the electrical analogue of diffusion, it is apparent that a high-diffusivity path corresponds to one of high conductivity, and this will tend to relieve the potential gradient or act like a short circuit. However, this analogy is not immediately apparent to many, and the phrase "high-diffusivity path" is used here instead.

boundaries double the measured diffusion coefficient.   If one-millionth of the atoms make a contribution to the flux which is comparable to that of all the rest of the atoms, then each of these atoms must be jumping roughly one million times as often as the regular lattice atoms. At lower temperatures the difference in the jump frequencies for the grain boundaries and in the grains would be even larger.   Since the grain boundary atoms represent such a small part of the specimen, it also follows that the mean jump frequency in this region can be a few orders of magnitude larger than it is in the lattice, for example 10^3 times, and still the boundary regions will make no significant addition to the total flux.

The first problem to be dealt with is how to measure the diffusion coefficient in these high-diffusivity paths.   These paths cannot exist except as regions in otherwise perfect crystals.   It is apparent that D therein cannot be measured on a sample consisting only of such paths, so some means must be found of treating measurements made on samples in which the surface atoms represent a very small fraction of the atoms in the sample.   Two types of such solutions will be discussed.   The first of these uses a tracer concentration gradient as a driving force and the radioactivity of the tracer to measure the total amount of material transported. The second uses surface tension as a driving force and obtains the total flux from the change in the surface contour of the sample.

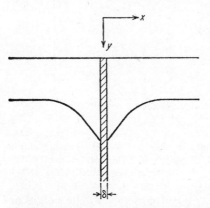

FIG. 6-2. Coordinate axes and isocon-centration line in a section of the model used for grain boundary diffusion analysis.

6-1. ANALYSIS OF GRAIN BOUNDARY DIFFUSION

To obtain values of the grain boundary diffusion coefficient D_b from diffusion studies on bicrystals, Fisher[1] suggested the following analysis.   Consider the grain boundary to be a thin layer of high-diffusivity material between two grains which have a low diffusivity.   A section normal to the grain boundary and the free surface is shown in Fig. 6-2.   A layer of tracer is applied to the free surface, that is, the plane $y = 0$, and allowed to diffuse into the sample.   The problem is to determine the concentration $c(x,y,t)$.   The

[1] J. C. Fisher, *J. Appl. Phys.*, **22**: 74 (1951).

boundary conditions are:

$$c = c_0 \quad \text{for } y = 0 \text{ and } t \geq 0$$
$$c = 0 \quad \text{for } y > 0, \text{ at } t = 0$$

It is further assumed that there is no concentration gradient across the grain boundary and that the concentration varies continuously in going from the grain boundary slab to the grain.

To obtain the differential equation which is valid inside the high-diffusivity slab, consider an element of this slab which is dy long by δ thick by unit length deep (into the sheet in Fig. 6-2). The fluxes into, or out of, the faces normal to the x and y axes are shown in Fig. 6-3. Any plane normal to the z axis is a symmetry plane, so J_z would equal zero. A dimensional argument shows

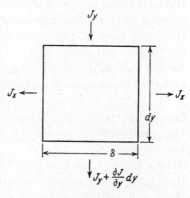

FIG. 6-3. Fluxes into and out of an element of grain boundary slab.

that $\partial c/\partial t$ for this element is given by an equation of the form

$$\frac{\partial c}{\partial t} = \frac{1}{1\, dy\, \delta}\left[\delta\left(J_y - J_y - \frac{\partial J_y}{\partial y}\, dy\right) - 2\, dy\, J_x\right]$$

$$\frac{\partial c}{\partial t} = \frac{-\partial J_y}{\partial y} - \frac{2}{\delta} J_x \tag{6-1}$$

J_x is the flux out of the grain boundary into the perfect lattice and can be replaced by $-D_l(\partial c/\partial x)$ with the gradient evaluated in the lattice just outside the slab. An expression for J_y can be obtained if a grain boundary diffusion coefficient D_b is defined by the equation

$$J_y = -D_b \frac{\partial c}{\partial y} \tag{6-2}$$

Substituting in Eq. (6-1) then gives

$$\frac{\partial c}{\partial t} = D_b \frac{\partial^2 c}{\partial y^2} + \frac{2D_l}{\delta}\left(\frac{\partial c}{\partial x}\right)_{x=\delta/2} \tag{6-3}$$

Outside the grain boundary, diffusion would obey the equation

$$\frac{\partial c}{\partial t} = D_l \nabla^2 c \tag{6-4}$$

The problem thus becomes one of determining the solution which will simultaneously satisfy these differential equations in the respective regions and be continuous, or "match up," across the boundary between the slab and the grain.

Solutions to Grain Boundary Diffusion Problem.   There have been three solutions given for the above problem.[1,2]   The complexity of each exceeds that of the preceding; only Fisher's relatively simple, and relatively inexact, procedure will be given here.   This is adequate for our present purpose.   The interested reader is referred to Whipple and Levine and MacCallum for a discussion of the appropriate corrections and the extension to a polycrystal sample with nonplanar boundaries.

In working with numerical solutions to Eqs. (6-3) and (6-4), Fisher found that the concentration in the grain boundary rose quickly at first but continued to rise at an ever-decreasing rate.   Thus the grain boundary concentration at any point, $c_b(y,t)$, will be near its final value during much of the anneal.   To simplify the solution, he assumed that the grain boundary concentration at each depth stayed at some constant value during the entire anneal.   It was also assumed that the flux of solute in the grains was normal to the grain boundary slab.   These assumptions allowed him to replace the initial system with a series of slices normal to the y axis, of thickness dy.   These are separated by impermeable membranes (so that all flow in the lattice will be normal to the slab).   At time $t = 0$, some concentration $c_b(y)$ is imposed on each slice at the slab and held constant for the duration of the experiment.   The concentration in each slice is then given by the equation

$$c(x,y,t) = c_b(y)\left[1 - \text{erf}\left(\frac{x}{2\sqrt{D_l t}}\right)\right] \tag{6-5}$$

The concentration contours assumed to exist in the actual "unsliced" specimen are then obtained by connecting the points of equal concentration in this "sliced" model.   This gives

$$c(x,y,t) = c_0 \exp\left[\frac{-y\sqrt{2}}{(\pi D_l t)^{\frac{1}{4}}(\delta D_b/D_l)^{\frac{1}{2}}}\right]\left[1 - \text{erf}\left(\frac{x}{2\sqrt{D_l t}}\right)\right] \tag{6-6}$$

Experimentally it is easiest to determine the amount of tracer in each of a series of slices dy thick and parallel to the free surface. This amount will be just that quantity of solute drained off from the

[1] J. C. Fisher, *J. Appl. Phys.*, **22**: 74 (1951).

[2] R. T. Whipple, *Phil. Mag.*, **45**: 1225 (1954).   H. S. Levine and C. J. MacCallum, *J. Appl. Phys.*, **31**: 595 (1960).

grain boundary by each slice or

$$\bar{c}(y,t) \, dy = c_b(y) \, dy \int_{-\infty}^{\infty} \left[1 - \mathrm{erf}\left(\frac{x}{2 \sqrt{D_l t}} \right) \right] dx$$
$$= c_b(y) \, dy \text{ (const)} \tag{6-7}$$

A plot of ln $\bar{c}$ versus y then should give a straight line of slope

$$\frac{- \sqrt{2}}{(\pi D_l t)^{\frac{1}{4}} (\delta D_b / D_l)^{\frac{1}{2}}} \tag{6-8}$$

This solution would be valid only in the region of y greater than, say, $4 \sqrt{D_l t}$, where most of the tracer present has entered by diffusing out from the grain boundary instead of diffusing in the matrix parallel to the boundary. Experimentally this is observed if the ratio D_b/D_l is large enough, but the Fisher analysis gives no indication of when D_b/D_l is "large enough." For an indication of this, it is necessary to go to one of the more complicated analyses. Figure 6-4 is adapted from Whipple and shows concentration contours for $c = 0.2c_0$ and for $\beta = 0.1$, 1.0, and 10 where

$$\beta = \frac{D_b \delta}{2 D_l \sqrt{D_l t}}$$

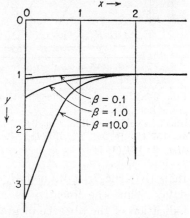

FIG. 6-4. Concentration contours for $c = 0.2c_0$ showing degree of extra penetration near grain boundary for various values of β. Plotted in units of penetration in absence of grain boundary. [*After R. Whipple, Phil. Mag.*, **45**: 1225 (1954).]

It is seen that there is no significant additional penetration along the grain boundary until $\beta \geq 1$. For $\delta = 4 \times 10^{-8}$ cm, $D_l = 10^{-11}$ cm²/sec, and $t = 10^5$ sec ($\simeq$28 hr); this means that D_b/D_l must be greater than 5×10^4 before there is appreciable penetration at the grain boundaries. Physically the reason for this is that the grain boundary slab is so thin that the flux along it is not sufficient to bring in enough material to distort the contours until $D_b/D_l > 5 \times 10^4$.

Experiments to check this analysis have been performed on bicrystals with plane grain boundaries. However, to show the effect of grain boundaries on the apparent values of D_l, we shall take as an example the work of Wajda[1] using polycrystalline zinc samples. The plane end of cylindrical samples was plated with radioactive zinc, the samples annealed, and cuts taken parallel to the plated face. A plot of activity (concentration) vs. distance is shown in Fig. 6-5. In this sample $D_b/D_l \simeq 7 \times 10^4$, and $\beta \simeq 3$, so that it was annealed in the temperature range where grain boundary diffusion is just becoming important. Several interesting conclusions can be seen from this work:

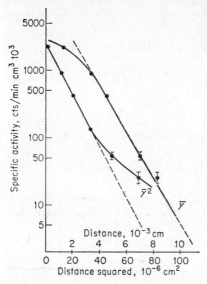

FIG. 6-5. Example of the effect of grain boundaries for $\beta = 3.2$. Zinc tracer in polycrystalline zinc, 312.5 hr at 157°C. [*From E. Wajda, Acta Met.*, **2**: 184 (1954).]

1. The experimental points in Fig. 6-5 do not deviate from a line of $\ln c \sim y^2$ until $c \simeq 0.1c_0$. This is consistent with Whipple's analysis which showed that for a fixed β the concentration contours (as in Fig. 6-4) become more pointed at the grain boundary as the concentration of the contour decreases.

2. In spite of the fact that the points fit the line $\ln c \sim y^2$, the apparent value of D_l calculated therefrom is a factor of 6.5 larger than the true D_l obtained by extrapolating the higher temperature data. Thus the linearity of a plot of $\ln c$ versus y^2 is a very poor indication of the absence of grain boundary effects.

3. At truly "low" temperatures, for example, 90°C for Zn, the points all fall on a plot of $\ln c$ versus y. In this range $D_b/D_l \simeq 1.7 \times 10^6$.[†] The value of β obtained with a value of D_l extrapolated from the high-temperature data is about 10^3.

Before leaving this analysis, it should be pointed out that though the physical situation is quite different, the mathematical analysis given above is also applicable to the case of surface diffusion. The slab in

[1] E. S. Wajda, *Acta Met.*, **2**: 184 (1954).

† This value is taken from the paper by Levine and MacCallum and is roughly a factor of 4 larger than that found by Wajda using Fisher's equation. The correction stems from the fact that the grain boundaries are not plane, nor are the grains infinite as assumed in Fisher's analysis.

Fig. 6-2 is a plane of symmetry so there will be no net flux across it. Thus if the half of the bicrystal to the left of the slab is removed, the high-diffusivity slab remains, but now it corresponds to a solid-vapor interface. In the derivation of Eq. (6-3), the only change required is to remove the factor of 2 since the volume element in Fig. 6-3 now loses material to the lattice on only one side.

6-2. EXPERIMENTAL RESULTS ON GRAIN BOUNDARY DIFFUSION

There is no accepted model of the grain boundary region for anything except low angle boundaries. In this region a dislocation model is generally accepted. However, there is still no quantitative theory of why, and by how much, D_b is greater than D_l even for a low angle boundary. The main progress made to date consists of several

TABLE 6-1. Values of D_0 and ΔH for Grain Boundary and Lattice Diffusion

Metal	$D_{0b}{}^a \left(\dfrac{cm^2}{sec}\right)$	$D_{0l} \left(\dfrac{cm^2}{sec}\right)$	$\Delta H_b \left(\dfrac{kcal}{mole}\right)$	$\Delta H_l \left(\dfrac{kcal}{mole}\right)$	Ref.
Ag	0.09	0.7	21.5	45	1, 2
Zn^b	0.14	0.4	14	23	3, 4
Cd	0.7	0.1	13	18.5	5
Fe	8.8^c	18	40	64	6

[1] Boundary data are in R. E. Hoffman and D. Turnbull, *J. Appl. Phys.*, **22**: 634 (1951).

[2] Lattice data are in L. Slifkin, D. Lazarus, and T. Tomizuka, *J. Appl. Phys.*, **23**: 1032 (1952).

[3] Boundary data are in E. S. Wajda, *Acta Met.*, **2**: 184 (1954).

[4] Lattice data are in G. Shirn, E. Wajda, and H. Huntington, *Acta Met.*, **1**: 513 (1953).

[5] E. Wajda, G. Shirn, and H. Huntington, *Acta Met.*, **3**: 39 (1955).

[6] C. Leymonie, Doctoral Thesis presented to Faculté des Sciences de L'Université de Paris, 1959.

[a] These values are obtained using $\delta = 3$ atom diameters. All data were obtained using Fisher's equations for a plane grain boundary, on polycrystalline samples. Thus the values of D_{0b} should probably be roughly a factor of 4 larger [see Levine and MacCallum, *J. Appl. Phys.*, **31**: 595 (1960)].

[b] In hcp zinc and cadmium, D_{0l} and ΔH_l for diffusion parallel and perpendicular to the c axis differ by 50 and 10% respectively. The values given here are average values.

[c] This is for Armco iron (99.7% Fe); D_{0b} for 99.96% Fe would appear to be one-fifth as large (Ref. 6).

systematic studies of the effect of temperature and grain boundary structure. These are summarized below.

Effect of Temperature. Figure 6-1 shows that grain boundary effects in polycrystalline silver become measurable only below 750°C and that they become relatively more important as the temperature is decreased. This is a general effect which is true for all metals studied and indicates that the activation energy for grain boundary diffusion is appreciably less than that for lattice diffusion. Some representative values of D_0 and ΔH for grain boundary and lattice diffusion are given in Table 6-1.

In a dislocation core or in any hard sphere model of a grain boundary, there will be many relatively open regions. In these regions the

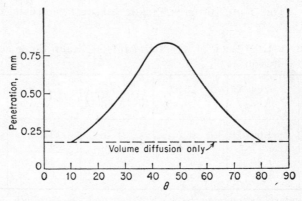

Fig. 6-6. Penetration of silver along (100) tilt boundaries in copper after two weeks at 675°C. [*After L. Couling and R. Smoluchowski, J. Appl. Phys.*, **25:** 1538 (1954).]

energy to form a vacancy or move an atom into a vacancy will be lower than in the lattice. Thus it is easy to see qualitatively why ΔH for self-diffusion in the boundary is less than in the lattice. However, no quantitative treatment has been given.

Effect of Grain Boundary Misorientation. The first systematic work on the variation of D_b with boundary misorientation was done by Smoluchowski and coworkers. They plated radioactive silver on a surface normal to the columnar axis of a copper specimen made up of columnar grains. After a diffusion anneal, material was removed parallel to the plated surface until the solute could not be detected in the grains. This depth was recorded, and polishing continued until an autoradiograph could no longer detect solute in the grain boundary region. Representative results are shown in Fig. 6-6 for symmetric

(100) tilt boundaries.[1]   The angle θ is the relative angle of rotation about the [100] direction required to bring the two crystals into coincidence.   It is seen that no preferential grain boundary penetration can be detected for $\theta \simeq 10°$ or $>80°$, but that for $\theta > 10°$ the penetration increases to a maximum at about $\theta = 45°$ and then decreases again to zero at $\theta \simeq 80°$.   The same sort of effect has been found for other metals and other types of tilt boundaries.[2]

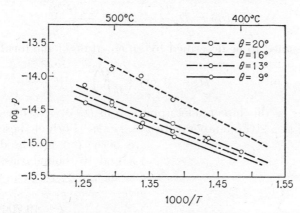

FIG. 6-7. Dependence of log $(D_b\delta)$ on temperature for various θ in (100) tilt boundaries in silver.   [*After D. Turnbull and R. Hoffman, Acta Met.*, **2**: 419 (1954).]

Hoffman and Turnbull[3] have made quantitative measurements of the ratio D_b/D_l for Ag^{110} in symmetric (100) tilt boundaries in silver in the range $9° \leq \theta \leq 28°$.   Their results are shown in Fig. 6-7, where p is defined equal to $D_b\delta$.   It is seen that ΔH is the same for the 9, 13, and 16° boundaries but that $D_0\delta$ increases with increasing θ. Their interpretation of these data involves a reexamination of the grain boundary diffusion model used by Fisher.

The dislocation model for low angle grain boundaries is now well established.   This model predicts that a low angle (100) tilt boundary consists of edge dislocations parallel to the [100] and a distance

---

[1] The two halves of a bicrystal containing a (100) tilt boundary have a common [100] direction in the boundary and can be brought into coincidence by rotation about the [100].   If the grain boundary also is a plane of symmetry, it is said to be a symmetric tilt boundary.   If the common [100] direction is normal to the boundary, the boundary is called a (100) twist boundary.

[2] For a recent review, see F. Weinberg, in Chalmers and King (eds.), "Progress in Metal Physics," vol. 8, p. 105, Pergamon Press, Inc., New York, 1959.

[3] D. Turnbull and R. Hoffman, *Acta Met.*, **2**: 419 (1954).

$d = a_0/2 \sin (\theta/2)$ apart.[1] The lattice between the dislocation cores is elastically strained but relatively perfect. Turnbull and Hoffman postulated that in the core of these dislocations the diffusion coefficient D_p is much greater than D_l. Thus instead of replacing the grain boundary by a slab of uniform thickness δ and diffusivity D_b, they replace it with a planar array of "pipes" of cross-sectional area h^2 and spacing d. For diffusion in the direction of the dislocation cores or pipes, this gives the equation

$$p = D_b\delta = D_p \frac{h^2}{d} = 2D_p h^2 \frac{\sin (\theta/2)}{a_0} \qquad (6\text{-}9)$$

If D_p is assumed to be described by an equation of the form

$$D_p = A \exp \left(\frac{-Q}{RT}\right)$$

then as long as this pipe model applies, $\partial \ln D_p/\partial(1/T)$ does not vary with θ, and at a given temperature $p \sim \sin \theta$. Both of these relations are borne out for the data on 9, 13, and 16° boundaries. Taking $h \simeq a_0 \simeq 5 \times 10^{-8}$ cm, their data give

$$D_p = 0.1 \exp \left(-\frac{19,700}{RT}\right) \text{cm}^2/\text{sec} \qquad (6\text{-}10)$$

It appears then that the absence of observable preferential grain boundary diffusion for $\theta \leq 10°$ in Fig. 6-6 is the result of a lower resolution technique and a higher temperature anneal.

Anisotropy of D_b in Tilt Boundaries. A corollary of the pipe model for diffusion is that in a given low angle boundary the excess diffusivity parallel to the cores ($p_\parallel$) should be greater than

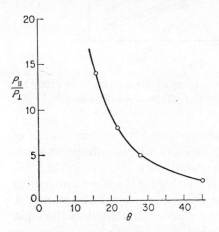

Fig. 6-8. Dependence of the anisotropy of grain boundary diffusion on θ for (100) tilt boundaries in silver. $T = 450°C$. [*From R. Hoffman, Acta Met.*, **4:** 98 (1956).]

in the direction perpendicular to the cores ($p_\perp$). This anisotropy, that is, $p_\parallel/p_\perp > 1$, should get more pronounced as θ decreases, while for higher values of θ, where the dislocation model breaks down, isotropy might be suspected. Figure 6-8 shows the results of Hoffman on (100)

[1] T. Read, "Dislocations in Crystals," chap. 11, McGraw-Hill Book Company, Inc., New York, 1953.

symmetric tilt boundaries in silver. Not only does the anisotropy decrease smoothly as θ increases, but it persists up into the region of high θ where the dislocation model is no longer applicable. Indeed anisotropy persisted up to $\theta = 45°$. Thus the assumption that high angle grain boundaries are homogeneous, isotropic slabs is seen not to be true, and this assumption should accordingly be used with caution.

6-3. DISLOCATION EFFECTS[1]

In a well-annealed metal single crystal any plane passed through the crystal will cut 10^6 to 10^7 dislocations per square centimeter. If a plane were passed perpendicular to the grain boundary in the 9, 13, or 16° (100) tilt bicrystals discussed above, it would cut about 10^7 dislocations per square centimeter of grain boundary plane intersection. The primary difference between the two cases is that in the latter all of the dislocation lines are parallel, while in the former case the dislocations run in all directions. Thus the grain boundary increases the apparent diffusivity in the plane of the grain boundary while the randomly oriented dislocations in the single crystal increase the apparent diffusivity in all directions. The heat-flow analogy to the grain boundary diffusion problem was a system consisting of a sheet of aluminum foil between two sheets of plastic. For the annealed single crystal, the heat-flow analogy would be a system consisting of fine aluminum wires randomly distributed in plastic.

We wish to relate the apparent, or macroscopic, diffusion coefficient to D_l and D_p. This could be done by using Fick's first law and the heat-flow analogy. However, we shall instead use a random-walk analysis since the extension from pure metals to solute effects is easier to treat.

Assume that each atom in a set makes n jumps in a pure metal single crystal containing many randomly oriented dislocations. The net displacement for each atom after n jumps is R_n. Ignoring correlation effects and assuming all jump distances to be r units long, the average of the various R_n^2 terms is given by the equation

$$\overline{R_n^2} = nr^2 \tag{6-11}$$

Now of the n jumps taken by an atom, n_l were in the perfect lattice, and n_d were in the dislocation pipe. In averaging over many atoms,

[1] E. Hart, *Acta Met.*, **5**: 597 (1957).

Eq. (6-11) can be written

$$\overline{R_n{}^2} = nr^2 = (\bar{n}_d + \bar{n}_l)r^2 \tag{6-12}$$

The jump frequencies inside and outside the dislocations[1] can be defined by the equations

$$\bar{t}_d\Gamma_d = \bar{n}_d \qquad \bar{t}_l\Gamma_l = \bar{n}_l$$

where $\bar{t}_d$ and $\bar{t}_l$ are the average times spent on and off, respectively, of the high-diffusivity sites around the dislocation core. Substituting these terms in Eq. (6-12) and dividing by the average time to make n jumps, we have

$$\frac{\overline{R^2}}{\bar{t}} = \Gamma_d r^2 \frac{\bar{t}_d}{\bar{t}} + \Gamma_l r^2 \frac{\bar{t}_l}{\bar{t}}$$

Except for a geometric constant, $\overline{R^2}/\bar{t}$ equals the apparent diffusivity D; and except for the same constant, $\Gamma_d r^2$ and $\Gamma_l r^2$ equal D_p and D_l, respectively. Thus the apparent diffusion coefficient for a single crystal is

$$D = D_p g + D_l(1 - g) \tag{6-13}$$

where $g = \bar{t}_d/\bar{t}$.

For the case of a pure metal, e.g., radioactive silver in silver, all atoms are equivalent, and the fraction of time a tracer spends "inside" a dislocation is just equal to the fraction of the total number of sites that are inside a dislocation. Taking the density of dislocations to be 10^7 per cm² (as would be expected for a not too carefully prepared single crystal) and the number of atoms per plane "inside" a dislocation to be 10, we have

$$g = \left(\frac{10^7 \text{ disl.}}{\text{cm}^2}\right)\left(\frac{\text{cm}^2}{10^{15} \text{ atoms}}\right)\left(\frac{10 \text{ atoms}}{\text{disl.}}\right) = 10^{-7}$$

Since g is so small, Eq. (6-13) can be rewritten

$$D = D_p g + D_l$$

or
$$D/D_l = 1 + (D_p/D_l)g \tag{6-14}$$

---

[1] As in the case of grain boundary diffusion, there will be a range of values of Γ or D for atoms as their distance from the core of the dislocation increases. The concepts of "inside" and "outside" the dislocation, and dislocation "pipe," are only introduced to make the calculations easier. The reader should avoid the error of taking these terms too literally.

Using Eq. (6-10) for D_p in silver and taking D_l from Table 6-1 gives

$$D_p/D_l \simeq 0.15 \exp\left(+\frac{25,000}{RT}\right) \tag{6-15}$$

At 500°C this gives $D_p/D_l \simeq 0.2 \times 10^7$. Substituting in Eq. (6-14) gives $D/D_l \simeq 1.2$. Returning to Fig. 6-1, it is seen that at 500°C the measured points for single crystals lie at least a factor of 2 above the line extrapolated from the higher temperature points. The deviation of these low-temperature points is in the direction predicted by the above analysis, and the observed increase could be explained with the reasonable value of 10^8 dislocations per square centimeter. A more careful comparison of this sort has been attempted by Tomizuka,[1] but there are too many uncertainties in the data, the analysis, and the system to allow a really precise comparison.

TABLE 6-2. Values of $D_p g/D_l$ at Various Temperatures for Ag with $g = 10^{-7}$

$D_p g/D_l$	$T(°C)$	T/T_{mp}
0.08	590	0.7
0.90	465	0.6
27	345	0.5
4500	220	0.4

The temperature dependence of D_p/D_l is given in Eq. (6-15). To emphasize how important the dislocation contribution becomes at lower temperatures, consider the values given in Table 6-2. Here g is taken to be 10^{-7}, and the data are strictly true only for silver. However, the values of $D_p g/D_l$ should be roughly the same for other metals at comparable values of the ratio of the temperature of the experiment to the temperature of the melting point (T/T_{mp}).[2] Thus for temperatures at or below half the melting point, the apparent lattice diffusion coefficient is entirely determined by the dislocation density, and the values of D_l extrapolated from higher temperatures are of little or no value in determining how rapidly matter is transported.

The contribution of dislocations to the apparent value of D in pure metals has several important ramifications. First, it sets a definite limit on the temperature range over which one can work in an effort to make a very accurate determination of ΔH for self-diffusion using

[1] C. Tomizuka, *Acta Met.*, **6**: 600 (1958).
[2] Such fractions are always determined using the absolute temperature scale.

tracers. If the values of D are determined more accurately, the allowable contribution of dislocation pipes is lower. However, the only way to decrease the effect of dislocations (for a given dislocation density) is to raise the temperature. Since the melting point sets an upper limit on the temperature range, the net effect is to decrease the temperature range over which D can be reliably measured.

A similar difficulty arises in attempting to make an accurate determination of ΔH for grain boundary diffusion. It was pointed out above that D_b/D_l has to be 10^5 or greater before the grain boundaries have enough effect to allow the accurate determination of $D_b\delta$. Roughly speaking, this means that the temperature of the measurement must be less than $\frac{2}{3}T_m$. However, when $T \leq \frac{1}{2}T_m$, the uncertainty in D_l due to dislocation effects makes it impossible to determine $D_b\delta$ from the experimentally determined quantity $D_b\delta/D_l$. Thus the usable temperature range is relatively short, and the resulting values of ΔH_b are not of high accuracy.

Solute Diffusion Down Dislocations. The differences between solute and solvent diffusion down dislocations have not been clarified. However, one of the differences results from the interaction which can exist between dislocations and solutes. If the solute atom is attracted to the dislocation, the fraction of the time a solute spends inside a dislocation is greater than the fraction of the sites inside dislocations. Mortlock[1] has suggested that the fraction of the time inside dislocations (g) can be approximated by the expression $(c_\perp/c_0)f$ where $c_\perp$ is the equilibrium concentration inside a dislocation, c_0 is the solute concentration in the perfect lattice, and f is the fraction of sites inside dislocations. The ratio of the apparent solute diffusion coefficient to the true diffusion coefficient for the solute thus becomes

$$D/D_l = 1 + (D_p/D_l)f(c_\perp/c_0) \tag{6-16}$$

The effect of this interaction can be appreciable. As an example, in the cases of iron and cobalt in copper, Mortlock concludes that $c_\perp/c_0$ is between 10 and 100.

As an example of other, less well-understood, effects which can arise in this area, Ainslie, Phillips, and Turnbull[2] have shown that sulphur diffusing into α-Fe from the grain boundaries actually generates a network of new dislocations, and that the sulphur then segregates to the dislocations. Any attempt to analyze these data in terms of simply D_b, D_l, and δ could be extremely misleading.

[1] A. Mortlock, *Acta Met.*, **8**: 132 (1960).
[2] N. Ainslie, V. Phillips, and D. Turnbull, *Acta Met.*, **8**: 528 (1960).

6-4. DIFFUSION DRIVEN BY SURFACE TENSION

There is a variety of diffusion-controlled phenomena familiar to a metallurgist, which are driven by the surface tension or surface free energy of solids. For example, in the sintering of pure metal powders the only driving force is the decrease in free energy which accompanies the great reduction in the vapor-solid interfacial area. Other examples are the spheroidization of pearlite and the coarsening of precipitate particles in an age-hardening alloy. Here there is a decrease in the interfacial area between the two-solid phases. The principles which would go into calculating the rate of each of these processes are understood. However, in each case the actual calculation of the rate would be rendered very difficult by the complicated geometry.

The only example of a surface-tension-driven reaction which will be discussed in detail is the case of grain boundary grooving on pure metal surfaces. The reasons for this are:

1. A rigorous mathematical analysis exists.
2. The analysis is borne out by experiment.
3. The experiments yield values of the surface self-diffusion coefficient.
4. The analysis points up the various transport mechanisms which can contribute in surface-tension-driven processes.

Analysis of Grain Boundary Grooving. The system to be analyzed consists of a pure metal bicrystal with a plane grain boundary which is normal to the surface. A section normal to the grain boundary and the surface is shown in Fig. 6-9a. There is a free energy or surface tension associated with each surface, which tends to decrease the area of that surface.[1] Thus at the intersection of the grain boundary and the solid-vapor surface, the grain boundary will shorten in the y direction until the surface tension of the grain boundary γ_b is just counteracted by the surface tension of the two solid-vapor surfaces γ_s. This local equilibrium at the intersection is attained when

$$\gamma_b = 2\gamma_s \sin \theta \qquad (6\text{-}17)$$

where θ is the angle between the free surface at the root of the groove and the free surface far from the groove (see Fig. 6-9a).

---

[1] In a crystalline solid the free energy of a surface may depend on its orientation relative to the lattice. However, in the analysis given here, it is assumed that γ_s is the same for all orientations exposed in the groove.

The groove in the otherwise flat surface gives rise to a **curvature in** the surface, but the atoms in a surface of radius of curvature r **have a higher chemical potential** than the atoms in a flat surface at the **same**

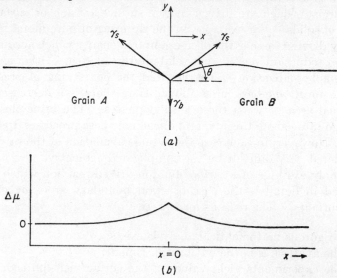

(a)

(b)

FIG. 6-9. (a) Section of bicrystal normal to surface, and grain boundary **showing** grain boundary groove. (b) Chemical potential of atoms on surface **as a function** of position along the surface shown directly above it.

pressure and temperature. This difference in chemical **potential is** given by the Gibbs-Thomson equation as

$$\Delta\mu = \frac{\gamma\Omega}{r} = \gamma\Omega K \tag{6-18}$$

where Ω is the volume per atom and K is the **curvature of the surface.** The precise relationship for K is

$$K = \frac{d^2y/dx^2}{[1 - (dy/dx)^2]^{\frac{3}{2}}}$$

but since γ_b/γ_s is about $\frac{1}{3}$, θ is small, and $(dy/dx)^2$ will be **much less** than 1. Under these conditions $\Delta\mu$ can be approximated by **the** equation

$$\Delta\mu = \gamma\Omega \frac{d^2y}{dx^2} \tag{6-19}$$

Directly below Fig. 6-9a is shown the chemical potential of **the** **atoms** in the surface at each value of x. It is seen that $\Delta\mu$ is largest at $x = 0$ and decreases on either side. This decrease in $\Delta\mu$ with x is a gradient in μ and will give rise to a flux of surface atoms away from the grain boundary on both sides. These fluxes will widen and deepen the groove.

To determine the rate at which this enlarging occurs, one must set up a flux equation. However, a moment's reflection shows that flow can occur by diffusion along the surface, through the crystal, or through the vapor. To estimate the relative contributions of these three paths, consider the following argument. The flux carried by each path in the x direction is given by the product of the average velocity of the atoms per unit force along that path (D_i/kT), the

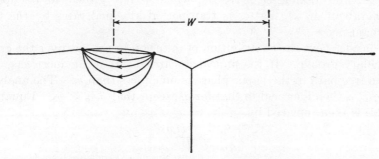

Fig. 6-10. Schematic drawing indicating the multiplicity of paths available for atoms going from one point to another through the solid.

density of atoms on the path ρ_i, and the average force per atom $-\partial\mu/\partial x$. Thus the general flux equation is

$$J_i = \frac{-D_i \rho_i}{kT} \frac{\partial\mu}{\partial x} \qquad (6\text{-}20)$$

The total quantity of material carried by each type of path in unit time is the flux times the effective cross-sectional area for that path. The relative contributions of surface and lattice diffusion are then given by

$$\frac{q_s}{q_l} = \frac{D_s}{D_l} \frac{\rho_s}{\rho_l} \frac{A_s}{A_l} \frac{(\partial\mu/\partial x)_s}{(\partial\mu/\partial x)_l} \qquad (6\text{-}21)$$

If we take the density of the surface material to be equal to the density of the bulk material, then $\rho_s/\rho_l = 1$. Surface diffusion will occur in a region about a_0 deep (corresponding to δ in the case of grain boundary diffusion), so that taking a unit length parallel to the groove gives $A_s \simeq a_0$. In the case of lattice diffusion, there are a great number of available paths between any two points on the surface. Some of these are shown schematically on Fig. 6-10. As the paths go deeper into the crystal, they make less of a contribution to q_l than those nearer the surface. As an approximation to the actual situation, we shall assume that J_l is the same for all paths down to a depth of the groove width w, and that for depths greater than w, J_l equals zero. Taking $(\partial\mu/\partial x)_l = (\partial\mu/\partial x)_s$ for all these paths and $A_l = w$ gives

$$\frac{q_s}{q_l} = \frac{D_s a_0}{D_l w} \tag{6-22}$$

If there is initially no groove in the surface, then $w \simeq a_0$, and since $D_s \gg D_l$, the result is $q_s/q_l \gg 1$. However, w increases as diffusion occurs; and taking $D_s \simeq 10^5 D_l$, when w has grown to be about $10^5 a_0$ (about 40 μ), matter is transported at equal rates by the two mechanisms.

Consider next the contribution of diffusion in the gas over the grain boundary groove. If the anneal is carried out in an inert gas, the mean free path in the vapor phase is much less than w. The analysis for q_{vap} is then identical to that for q_l except that $\rho_{vap} \ll \rho_l$. Equation (6-22) is then replaced by

$$\frac{q_s}{q_{vap}} \simeq \frac{D_s a_0 \rho_l}{D_v w \rho_{vap}}$$

Again for very small widths surface diffusion is rate controlling, while for larger widths vapor diffusion or lattice diffusion may be the dominant mechanism, depending on the ratio $(D_{vap}\rho_{vap})/(D_l\rho_l)$. A more complete and more rigorous discussion of the relative contributions of these various paths has been given by Mullins.[1]

We turn now to the calculation of the rate at which a grain boundary groove enlarges when all matter transport occurs by surface diffusion. This problem has been treated by Mullins,[2] and the following analysis is essentially that given by him. For surface diffusion, the flux equation in atoms per square centimeter per second is given by Eq. (6-20) as

$$J_s = -\frac{D_s}{kT}\frac{1}{\Omega}\frac{\partial\mu}{\partial x}$$

where ρ_s has been replaced by the reciprocal of the atomic volume Ω^{-1}. This equation is analogous to Eq. (6-2) which defined D_b. If we consider an element of surface area dx wide and having unit length in the direction of the groove root, the accumulation of material in this

[1] W. W. Mullins, *J. Appl. Phys.*, **30**: 77 (1959).
[2] W. W. Mullins, *J. Appl. Phys.*, **28**: 333 (1957).

element is proportional to[1]

$$dJ = J_s - J_{s+ds} = -\frac{\partial J}{\partial x}\,ds = \frac{D_s\partial^2\mu}{kT\Omega\partial x^2} \tag{6-23}$$

This accumulation raises the surface, so dJ/dx is proportional to the rate of rise of the surface $(\partial y/\partial t)$. To obtain an equality instead of a proportionality, it is necessary to assume an effective depth for the high-diffusivity surface layer as was done in the case of grain boundary diffusion. If this depth is defined as δ, the number of atoms per second entering the element of surface area is $dJ\,\delta\,1$, and the rate of rise of the surface element is the volume accumulated per second over the area, or

$$\frac{(dJ\,\delta\,1)}{dx\,1} = \frac{\partial y}{\partial t} = \frac{D_s\delta}{kT}\frac{\partial^2\mu}{\partial x^2} \tag{6-24}$$

Using Eq. (6-19), this can be rewritten

$$\frac{\partial y}{\partial t} = \frac{D_s\delta\gamma_s\Omega}{kT}\frac{\partial^4 y}{\partial x^4} \tag{6-25}$$

For the case of grain boundary grooving, the boundary conditions taken by Mullins were:

1. The surface is initially flat, that is, $y(x,t) = 0$ at $t = 0$.
2. The equilibrium groove angle, $\theta = $ arc sin $(\gamma_b/2\gamma_s)$, is constant for $x = 0$, $t \geq 0$.
3. $(\partial^3 y/\partial x^3)_{x=0} = 0$ for all t; that is, there is no flux of atoms out of the grain boundary.

The solution to Eq. (6-25) for these boundary conditions is of the form

$$y(x,t) = (\tan\theta)(Bt)^{\frac{1}{4}}Z\left[\frac{x}{(Bt)^{\frac{1}{4}}}\right] \tag{6-26}$$

where $Z[x/(Bt)^{\frac{1}{4}}]$ is a power series, B is given by the equation

$$B = \frac{D_s\delta\gamma_s\Omega}{kT} \tag{6-27}$$

and $\tan\theta \simeq \sin\theta = \gamma_b/2\gamma_s$.[†] Two properties indicated by the form of this solution are: (1) the shape of the predicted profile is independent of

---

[1] It is assumed here that D_s does not vary with position along the surface.

[†] Mullins' expression is $B = D_s\nu\Omega^2\gamma_s/kT$. The difference is in his use of the term $\nu = \delta/\Omega$ where he calls ν "the number of atoms per unit area."

the time; and (2) all of the linear dimensions of the profile are proportional to $t^{\frac{1}{4}}$.

The solid curve in Fig. 6-11 shows the shape of the curve. The fact that one curve serves for all times means that the x and y coordinates of any given point on the curve increase as $t^{\frac{1}{4}}$. In particular, the distance between the two maxima on either side of the boundary (w)

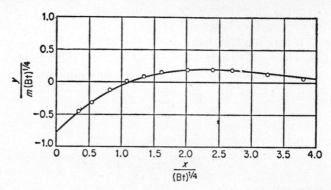

FIG. 6-11. Profile of surface formed in grain boundary grooving when surface diffusion is dominant. The curve is Eq. (6-26) and the points are the experimental data for copper. [*From W. Mullins and P. Shewmon, Acta Met.*, **7**: 163 (1959).]

and the depth from the maxima to the groove root (d) both increase as $t^{\frac{1}{4}}$. The equations for these are

$$w = 4.6(Bt)^{\frac{1}{4}} \tag{6-28}$$
$$d = 0.973 \tan \theta (Bt)^{\frac{1}{4}} \tag{6-29}$$

The humps on either side of the grain boundary result from the fact that no material leaves the region $|x| < 2w$. Thus all of the metal comes from beneath the original surface at the grain boundary and goes into the broad humps on either side of the boundary. The same type of humps would appear if diffusion in the solid or the gas were the dominant transport mechanism, provided there was no loss of material. However, in these cases the width between the humps would increase as $t^{\frac{1}{3}}$ instead of $t^{\frac{1}{4}}$.

6-5. DETERMINATION OF D_s FROM GRAIN BOUNDARY GROOVING

The quantities which appear in the expression for B [Eq. (6-27)] are all known for copper, except for $D_s\delta$. Thus if the shape of the grooves and their enlargement with time are those given by Eq. (6-26), the value of $D_s\delta$ can be obtained from the measured value of B. The points in Fig. 6-11 are for copper annealed in dry hydrogen and show that the shape of the groove can be satisfactorily described by Eq. (6-26).

The time dependence of the groove width w is shown in Fig. 6-12. The number written on each line is its slope and thus the exponent on t. The data show that, at higher temperatures and wider grooves, the

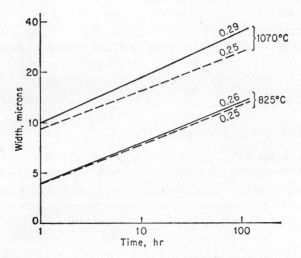

Fɪɢ. 6-12. Width of grain boundary grooves on copper vs. time for various temperatures. For each temperature there is a solid line representing the observed widths and a dashed line showing the contribution of surface diffusion. The number on each line is its slope. [*From N. Gjostein, Trans. AIME,* **221**: 1039 (1961).]

slope of the best line through the points lies above 0.25. However, after the contribution of volume diffusion is subtracted out, the resulting lines (dashed lines in Fig. 6-12) all have a slope of essentially $\frac{1}{4}$. Taking $\delta - \Omega^{-\frac{1}{3}}$ or about equal to the atomic diameter, the equation obtained for D_s on copper is[1]

$$D_s = 2 \times 10^4 \exp\left(-\frac{49,000}{RT}\right) \text{ cm}^2/\text{sec}$$

The value of D_s varies no more than a factor of 3 over all the surfaces studied, so that this equation is a good approximation for all surface orientations.

[1] J. Choi and P. Shewmon, *Trans. AIME,* **224**: 589 (1962).

The activation energy for surface self-diffusion (ΔH_s) on copper is almost equal to that for lattice self-diffusion (ΔH_l). This large a value of ΔH_s may at first seem strange since there is no straining of the lattice involved in moving an atom on the surface. However, an atom jumping along the surface of a metal is well on its way to being an evaporated atom. Thus ΔH_s should be more closely related to the heat of vaporization of the metal than it is to ΔH_l. The heat of vaporization of copper is 80 kcal/mole. If it is assumed that the binding energy of a copper atom is proportional to its number of nearest neighbors, it can be shown that $\Delta H_s = \frac{2}{3}\Delta H_{vap}$ for an fcc lattice.[1] This relation predicts that ΔH_s for copper is about 53 kcal. The same model predicts that ΔH_s should be the same for all surface orientations. Both predictions are in reasonable agreement with the experimental results.

6-6. D_s FROM FIELD EMISSION STUDIES

Another rather different technique which has been successfully used to study surface diffusion involves the field emission microscope. In this instrument a very high field is applied to a sharply pointed metal wire (radius of curvature, 0.1 to 1μ). In a high vacuum, the tip then emits electrons which form an observable pattern on a phosphorescent screen. Two types of experiments have been performed using this technique.

First, the blunting of the metal tip at $T \simeq \frac{1}{2}T_m$ is surface-diffusion controlled, and from the rate of blunting D_s can be determined.[2] For tungsten

$$D_s = 4 \exp \frac{-72,000}{RT} \text{ cm}^2/\text{sec}$$

One of the primary advantages of this technique is that a clean surface can be obtained on reactive metals. The number of metals which can be conveniently worked with is rather limited, but it is limited by different factors than those which limit the use of the grain boundary grooving technique, so the procedures should tend to complement one another.

In the second type of experiment, a solute which is chemisorbed, for example, O_2 on W, is applied to one side of the metal tip. The solute changes the work function locally, and the pattern of emitted electrons is changed. The rate of spreading of the solute over the metal surface

---

[1] J. Choi and P. Shewmon, *Trans. AIME*, **224**: 589 (1962).
[2] J. P. Barbour, F. M. Charbonnier, et al., *Phys. Rev.*, **117**: 1452 (1960).

can then be followed by studying the changing pattern. From this, an activation energy for the surface diffusion of the solute can be obtained and observations made about the variation of the diffusion coefficient with the surface orientation.[1]

PROBLEMS

6-1. It is desired to use small electric motors at 150°C. In preliminary tests at this temperature, the motors fail because the small copper wire in the windings oxidizes. It is suggested that a thin uniform layer of silver on these wires would protect them. As a diffusion expert, you are to calculate the thickness of silver required to maintain a $\geq 99\%$ silver alloy on the surface for 5 years. The most accurate data you can find are a study of the diffusion of silver in copper between 750 and 1050°C. Extrapolating these data to 150°C, you find that a 1μ layer of silver will last for 100 years. Some skeptic has a 1μ layer of silver put on a plate of copper and finds that the silver completely diffuses into the copper over a weekend at 150°C.

(*a*) Why was the calculation of the rate at 150°C invalid?

(*b*) What kind of an experimental program would you undertake to determine how thick the silver must be to maintain $\geq 99\%$ silver at the surface for 5 years? (You have 1 month to run the tests.)

6-2. Given a pure metal single crystal containing randomly distributed dislocations, use Fick's first law to give the equation for the flux across a plane which is normal to a tracer concentration gradient. What assumption must be made about the gradient around the dislocations before Eq. (6-13) is obtained?

6-3. If the work is carefully done, the self-diffusion coefficient in iron can be measured at 700°C. If there are 10^8 dislocations per square centimeter, estimate their contribution to the observed diffusion coefficient at 700°C. List all assumptions made.

6-4. In the text it was stated that below two-thirds of the melting point, diffusion along dislocations makes a marked contribution to solute self-diffusion studies in very dilute alloy single crystals. It is quite probable that interstitial atoms also diffuse much more rapidly along dislocations than in the lattice. What justification is there for having neglected the effect of this enhanced diffusion along dislocations in deriving values of D from internal friction studies?

6-5. A dislocation can be considered to be a cylinder of uniform high diffusivity surrounded by a low-diffusivity matrix. Derive the differential equation for $\partial c/\partial t$ inside the dislocation pipe.

[1] Reviews of this type of work are given by R. Gomer and A. J. Becker in "Advances in Catalysis," vol. 7, pp. 93 and 135, Academic Press, Inc., New York, 1955.

chapter 7 THERMAL

DIFFUSION AND

ELECTROLYSIS IN SOLIDS

This final chapter discusses two topics which are less well understood than the previous material, but which provide new and different information on diffusion in solids. The first of these, thermal diffusion, considers the effect of a temperature gradient on the migration of solute in a solid. The second, electrolysis, considers the effect of an electric potential gradient on the migration of solute in a solid. In thermal diffusion, the redistribution of solute which occurs is completely analogous to the more widely studied redistribution of electrons which occurs when a solid is placed in a temperature gradient and gives rise to thermoelectric effects.

In the case of electrolysis, the solute redistribution is similar to that studied in relating the electrical conductivity and matter flow in ionic conductors. In this chapter, though, we discuss the occurrence of the effect in metals where essentially all of the electric current is carried by electrons. Thus one can think of the experiments as measuring either the transport number of the solute atoms in an alloy or the effect of a large flux of electrons on the diffusive jumps of the solute atoms.

The discussion will deal primarily with a phenomenological treatment of the two effects since this allows the similarities of the two to be emphasized. A physical model will also be discussed for each effect, which allows some of the available data to be rationalized.

188

7-1. THERMAL DIFFUSION

It is an empirical fact that, if a homogeneous two-component phase is placed in a temperature gradient, an unmixing occurs. That is, one component diffuses preferentially to the hot end, tending to enrich the hot region with that component. This effect is called thermal diffusion—or the Sorét effect, after a scientist who made early studies of the effect in liquids. To obtain a flux equation that will fit this empirical observation, an additional term which is proportional to the temperature gradient must be added to Fick's first law. Thus the flux can be described by the equation

$$J_1 = -D_1 \frac{\partial c_1}{\partial x} - \beta_1 \frac{dT}{dx} \tag{7-1}$$

This equation must be satisfactory in the limit of small temperature gradients, and until proven otherwise, β_1 is assumed independent of dT/dx. The constant β_1 will be negative or positive depending on whether the solute flux in a homogeneous bar is up or down the temperature gradient. If β_1 is nonzero, it indicates that the probability a solute atom will make its next jump up the temperature gradient is different from the probability that it will make its next jump down the temperature gradient. Phenomenologically Eq. (7-1) is similar to that used to describe the effect of an electric field on diffusion in nonmetals in Chap. 5. As in that case, the small biasing of the jump direction will change neither the jump mechanism nor the mean jump frequency at any given temperature. Thus β_1 in Eq. (7-1) will be proportional to D_1. The constant relating β_1 to D_1 can be written in several ways, but it is most common to take $\beta_1 = (D_1 Q_1^* c_1 / RT^2)$. Equation (7-1) can then be rewritten as

$$J_1 = -D_1 \left(\frac{\partial c_1}{\partial x} + \frac{Q_1^* c_1}{RT^2} \frac{dT}{dx} \right) \tag{7-2}$$

Here the experimentally determined parameter which describes the sign and magnitude of the thermal diffusion effect is Q_1^*. If D_1/RT is taken to be the mobility of component 1, $-(Q_1^*/T)\, dT/dx$ is the effective "force" exerted by the temperature gradient on each solute atom. Q_1^* is called the heat of transport of component 1; the rest of our discussion of thermal diffusion will be devoted to a study of its interpretation and measurement.

The name "heat of transport" for Q^* and the expression for Eq. (7-2) come from the phenomenological equations of irreversible thermodynamics (also sometimes called the thermodynamics of the steady state). To derive Eq. (7-2), consider a specimen of a dilute binary alloy in which the mobility of the solute is so much greater than that of the solvent that we can assume that only the solute atoms move. (A very dilute solution of carbon in α-iron satisfies this assumption, since at 800°C the interstitial carbon diffuses roughly 10^5 times as fast as the iron atoms.) For a system in which concentration gradients and temperature gradients exist, there will be a heat flux J_q and a solute flux J_1. The general phenomenological equations allow for interaction of the two gradients, and are thus

$$J_1 = -M_{11}\left(\frac{\partial \mu_1}{\partial x}\right)_T - \frac{M_{1q}}{T}\left(\frac{\partial T}{\partial x}\right) \tag{7-3}$$

$$J_q = -M_{q1}\left(\frac{\partial \mu_1}{\partial x}\right)_T - \frac{M_{qq}}{T}\left(\frac{\partial T}{\partial x}\right) \tag{7-4}$$

where the temperature gradient and the chemical potential gradient have been taken parallel to the x axis.[1] The various M_{ij} are constants of proportionality and $(\partial \mu_1/\partial x)_T$ is the chemical potential gradient at constant temperature. $\left(\frac{\partial \mu_1}{\partial x}\right)_T$ and $\left(\frac{\partial T}{\partial x}\right)\frac{1}{T}$ are called forces and are said to be conjugate to the fluxes J_1 and J_q. There are several different sets of conjugate forces that could be used; those used here are the simplest for the present discussion.

Equation (7-2) can be put in a form similar to Eq. (7-3) by taking $d\mu/d \ln c_1 = RT$.† Equation (7-2) can then be rewritten

$$J_1 = -\frac{D_1 c_1}{RT}\left[\left(\frac{\partial \mu_1}{\partial x}\right)_T + \frac{Q_1^*}{T}\frac{dT}{dx}\right] \tag{7-5}$$

Factoring M_{11} out of Eq. (7-3) gives for the same system

$$J_1 = -M_{11}\left[\left(\frac{\partial \mu_1}{\partial x}\right)_T + \frac{M_{1q}}{M_{11}T}\frac{dT}{dx}\right] \tag{7-6}$$

Equations (7-5) and (7-6) will be equal if we take $M_{11} = D_1 c_1/RT$ and $Q_1^* = M_{1q}/M_{11}$.

We now have the basis to give an interpretation of $Q_1^* = M_{1q}/M_{11}$. In the absence of a temperature gradient, that is, $\nabla T = 0$, Eqs. (7-3)

---

[1] S. R. de Groot, "Irreversible Thermodynamics," chap. 5, North-Holland Publishing Company, Amsterdam, 1951.

† This is true whenever the activity coefficient is independent of composition and will be valid in ideal or dilute solutions.

and (7-4) give

$$\frac{J_q}{J_1} = \frac{M_{q1}}{M_{11}}$$

but $M_{q1} = M_{1q}$ by the Onsager reciprocal relations.[1]   Thus

$$\left(\frac{J_q}{J_1}\right)_{\nabla T=0} = \frac{M_{1q}}{M_{11}} = Q_1^* \tag{7-7}$$

It is seen then that Q_1^* is the heat flux per unit flux of component 1 in the absence of a temperature gradient.   Thus if $Q_1^* > 0$, a heat flux parallel to J_1 will be generated by a solute flux; that is, to keep the region gaining solute atoms isothermal, heat must be removed from it. If $Q_1^* < 0$, J_q and J_1 are in opposite directions, and the region gaining solute atoms must receive heat to keep it isothermal.   Our initial aim was to study the effect of a temperature gradient on diffusion, but the discussion has turned to the tendency of a diffusion flux to set up a temperature gradient in an isothermal system.   This interrelation of the two effects stems from our ability to relate J_1 and J_q through the equation $M_{q1} = M_{1q}$.   More fundamentally, this stems from the theorem of microscopic reversibility.   We shall not go into this theorem here, but the conclusion of interest is that if a thermal flux sets up a matter flux in a homogeneous system, a matter flux in an isothermal system will set up a thermal flux.

One way to obtain a better understanding of, and equations for, Q^* is to attempt an atomic analysis of why a matter flux would set up a thermal flux.   This may lead to physical insight, but attempts to obtain equations for Q^* in this way end up expressing Q^* as an undetermined function of both thermodynamic quantities and the activation energy for a jump (ΔH_m).†

Another approach to the interpretation of Q^* is to set up a kinetic model for the atomic jump process and derive an expression for Q^* using this model.   This approach has the advantage of using a well-defined model whose assumptions are clearly visible.   A kinetic argument due to Wirtz proceeds as follows: It is known from the kinetic analysis of diffusion in isothermal specimens that some high-energy

---

[1] A good introduction to these relations is given by K. G. Denbigh, "The Thermodynamics of the Steady State," Methuen & Co., Ltd., London, 1951. They hold whenever conjugate forces and fluxes are used, i.e., whenever the product of the flux J_i and the force X_i has the units (temperature)(entropy)/ (time)(volume).

† For example, see R. Oriani, *J. Chem. Phys.*, **34:** 1773 (1961).

configuration with an excess molar energy equal to ΔH_m must be obtained before an atomic jump can occur. The probability that a given solute atom will be in this high-energy configuration at any instant is related to the temperature of the specimen through the factor $\exp(-\Delta H_m/RT)$. In the presence of a temperature gradient, the average temperatures of the original, intermediate, and final planes of the jumping solute will be different. As a result, the frequency with which the required high-energy configuration is established for a jump to the higher temperature side may differ infinitesimally from that for a jump to the lower temperature side. To obtain an equation for these two jump frequencies, assume that ΔH_m may consist of three parts: (1) that which must be given to the atoms on the original plane of the solute atom (H_0), (2) that which must be given to the atoms in the intermediate plane of the jump (H_i), and (3) that which is required to prepare the final plane for the jumping atom (H_f). If this is done, then the jump frequency for atoms making a jump up the temperature gradient will be proportional to the product

$$\exp\left(-\frac{H_0}{RT}\right)\exp\left[-\frac{H_i}{R(T+\frac{1}{2}\Delta T)}\right]\exp\left[-\frac{H_f}{R(T+\Delta T)}\right]$$

where the average temperature difference between the original and the final plane of the jumping solute is ΔT. The jump frequency for atoms jumping in the reverse direction between the same two planes will be proportional to:

$$\exp\left(-\frac{H_f}{RT}\right)\exp\left[-\frac{H_i}{R(T+\frac{1}{2}\Delta T)}\right]\exp\left[-\frac{H_0}{R(T+\Delta T)}\right]$$

The middle exponent in both of these equations is the same, so the ratio of the jump frequencies between the two planes is

$$\exp\frac{-H_0+H_f}{RT}\exp\frac{-H_f+H_0}{R(T+\Delta T)}$$

Now the condition for zero net flux of solute between two planes is that the average number of atoms jumping from the hotter plane to the colder plane in any given time increment is equal to the number of atoms making the reverse jump in the same period of time. If n_h and n_c are the number of atoms per unit area on the hotter and colder planes, respectively, the condition for zero net flux is

$$\frac{n_h}{n_c}=\exp\frac{-H_0+H_f}{RT}\exp\frac{+H_0-H_f}{RT+R\,\Delta T} \tag{7-8}$$

If Δn is defined such that $n_h = n_c + \Delta n$ and λ is the distance between the initial and the final planes, the left side of this equation can be written

$$\frac{n_h}{n_c} = 1 + \frac{\Delta n}{n_c} \simeq 1 + \frac{\lambda}{n}\frac{dn}{dx}$$

If the exponents in the right side of Eq. (7-8) are combined, it becomes

$$\frac{n_h}{n_c} = \exp\frac{(-H_0 + H_f)\,\Delta T}{RT^2(1 + \Delta T/T)} \tag{7-9}$$

but $\Delta T/T \ll 1$, so Eq. (7-9) can be simplified and expanded to give

$$1 + \frac{\lambda}{n}\frac{dn}{dx} = 1 + \frac{(H_f - H_0)\,\Delta T}{RT^2} = 1 + \frac{(H_f - H_0)\lambda}{RT^2}\frac{dT}{dx} \tag{7-10}$$

It follows from Eq. (7-10) that the steady-state solute distribution in a closed system is given by the expression

$$\frac{d\ln n}{dx} = -\frac{H_0 - H_f}{RT^2}\frac{dT}{dx}$$

To relate Q_1^* to $H_0 - H_f$, note that in the steady state J_1 is zero, and Eq. (7-2) gives

$$\frac{d\ln c_1}{dx} = -\frac{Q_1^*}{RT^2}\frac{dT}{dx} \tag{7-11}$$

The equation for Q^* resulting from Wirtz's model is therefore

$$Q^* = H_0 - H_f \tag{7-12}$$

Since H_0, H_i, and H_f were defined such that $\Delta H_m = H_0 + H_i + H_f$, Eq. (7-12) requires that $|Q^*| \le \Delta H_m$ but allows any value of Q^* in this range.

If the model leading to Eq. (7-12) is valid, measurements of Q^* will give information on the spatial distribution of the activation energy which is required before a jump can occur. In Chap. 2 the atomic mechanisms and atomic rearrangements required for an atomic jump were discussed. The most common situation was one in which the diffusing atom had to pass through a constriction on its way to a relatively open, new site. If the primary barrier to diffusion were the movement of constricting atoms out of the way so that the diffusing atom could pass, then most of ΔH_m would be located in the intermediate plane; ΔH_m would then be about equal to H_i, and Eq. (7-12) dictates that Q^* would be almost zero. On the other hand, if the main part of ΔH_m were required to make the diffusing atom execute violent enough oscillations to move it to the saddle point (the constriction always being relatively open), the result would be $\Delta H_m \simeq H_0$, and $Q^* \simeq H_0 \simeq \Delta H_m$. In this latter case the solute would tend to concentrate at the cold end. This result can be seen by noting that if $\Delta H_m \simeq H_0$, the jump frequency of the solute atoms on the hotter of two adjacent planes will always be greater than that of the solute on the colder plane.   Thus, if the number of atoms jumping from the cold plane to the hot plane

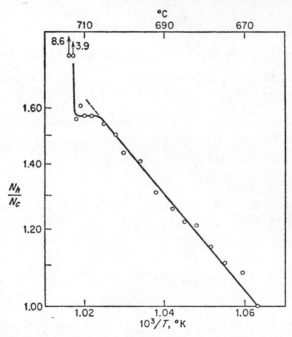

FIG. 7-1. Plot of carbon content (logarithmic scale) versus $1/T$ for α-iron specimen annealed in a temperature gradient until steady state is attained. [*From P. Shewmon, Acta Met.*, **8**: 606 (1960).]

per unit time is to equal the number making the reverse jump in the same time, there must be more solute atoms on the colder plane.

A system to which the above analysis might apply is carbon in iron. The results of a study on an initially single-phase iron-carbon alloy are shown in Fig. 7-1.   During the anneal the carbon concentration became higher at the hot end, indicating that Q^* is negative.[1]   The

[1] The off-scale points on the left side of Fig. 7-1 result from the precipitation of carbon in this region and in no way affect our discussion of $d \ln c/dT$ in the single-phase portion of the specimen.   For a general treatment of precipitation and the redistribution of a second phase during annealing in a temperature gradient, see P. Shewmon, *Trans. AIME*, **212**: 642 (1958).

line drawn through the points gives a value of $Q_c^*(\alpha) \simeq -23$ kcal. Experiments made on the same low-carbon (0.01% C) alloys in γ-iron gave $Q_c^*(\gamma) \simeq 0$.

These results can be rationalized in the following manner using Wirtz's model. In an fcc lattice an interstitial atom must pass through a pronounced constriction in a jump from one interstitial site to another. Thus it is plausible to say that H_i would be an appreciable part of ΔH_m and Q^* would be expected to be small, as it is. The large negative value of Q_c^* in the bcc α-iron may at first seem anomalous since its interpretation using Eq. (7-12) requires that most of ΔH_m goes into the plane of the final site of the jumping atom. However, a closer look at the bcc lattice, e.g., Fig. 7-2, shows that the moving atom must pass through no constriction midway along its jump. The major barrier to be overcome in the movement of an interstitial atom from one site to an adjoining one appears to be the moving apart of two adjacent iron atoms so that the carbon atom might jump from its initial position into a final position between them. The details of the case are not important enough to dwell on. The important point is that the model of Wirtz which led to Eq. (7-12) gives a physically understandable and apparently adequate

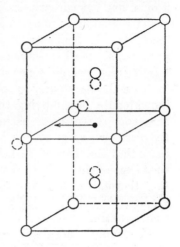

FIG. 7-2. The solid circles represent the position of the solvent atoms in a bcc lattice when an interstitial is at the position shown by the black circle (●). The dashed circles represent the final positions of the atoms that shift when the interstitial makes the jump shown by the arrow.

interpretation of the thermal diffusion effect. Actually, there is so little quantitative data on Q^* that it is impossible to honestly assess the adequacy of the model. This is illustrated by the fact that at present the only other interstitial solute studied is the case of hydrogen in zirconium. Here the Q^* is found to be $+6$ kcal/g atom in both the α phase (hcp)[1] and in the β phase (bcc).[2]

Several other types of experiments can be performed which involve thermal diffusion. In each, the analysis is more complicated than the example discussed above, though the basic flux equation for a given component [Eq. (7-2)] is unchanged. As an example, if a pure metal is

[1] A. Sawatsky, *J. Nuclear Material*, **2**: 321 (1960).

[2] J. Droege, *Battelle Mem. Inst. Rept.* 1502, February, 1961.

annealed in a temperature gradient, the concentration of vacancies at the hot end will be greater than at the cold end. A first impression might be that a flow of vacancies should occur down the vacancy concentration gradient, with an equal flux of atoms in the opposite direction. However, Eq. (7-2) indicates that the flux will also depend on the heat of transport for the vacancies. Vacancies, unlike solute atoms, are not conserved. If solute atoms flow from the hot to the cooler region of an alloy, the solute concentration in the cooler region increases. If vacancies flow from the hotter to the cooler region of a pure metal, the excess vacancies are destroyed in the cooler region, keeping the vacancy concentration unchanged. If markers are placed in or on the pure metal specimen, the destruction of vacancies will be indicated by the moving together of the markers. Experiments of this type have been performed using zinc,[1] iron,[2] and gold.[3] In the first two cases no marker movement was observed. This indicates that the thermal diffusion effect (due to $Q_v^* \neq 0$) just cancels out the effect of the vacancy concentration gradient. In the case of gold, the thermal diffusion effect opposes the vacancy concentration gradient, but there is still a net flux of vacancies from the hotter to the colder region.

An example of a still more complicated system is the thermal diffusion of silver in silver bromide. Since in AgBr all of the charge is carried by silver ions, the concentration difference of silver ions between the hot and the cold end sets up an electrical potential difference which can be measured with a potentiometer. Such a potential difference is more commonly related to the thermoelectric properties of the solid, but in the case of ionic solids it gives a direct indication of the equilibrium distribution of silver ions. The silver can move by either an interstitialcy mechanism or a vacancy mechanism. Since these defects have different mobilities and different values of Q^*, the equations are more complex than those discussed above.[4]

7-2. ELECTROLYSIS OF SOLIDS

If an iron-carbon alloy is heated to a temperature which makes it entirely austenitic and a direct current of electricity is passed through

[1] P. Shewmon, *J. Chem. Phys.*, **29**: 1032 (1958).

[2] W. Brammer, *Acta Met.*, **8**: 630 (1960).

[3] C. J. Meechan and C. Lehman, *J. App. Phys.*, **33**: 634 (1962).

[4] The equations relating Q^* and the thermoelectric power for ionic conductors are derived and discussed by Lidiard in "Handbuch der Physik," vol. 20, Springer-Verlag, Berlin, 1957.

it, there is a flux of carbon toward the negative electrode (cathode). The occurrence of such an effect in a liquid electrolyte or an ionic solid is common knowledge; its occurrence in metals is not nearly as well known, though it appears to occur in most, or all, alloy systems.

For simplicity, we again restrict ourselves to the case of an alloy with an interstitial solute. The phenomenological treatment of the electrolysis of such an alloy is completely analogous to that given above for thermal diffusion. There will be two fluxes, the flux of solute (J_1), and the flux of electric charge (J_e). The force conjugate to J_e will be $\partial\phi/\partial x$, so the flux equations are

$$J_1 = -M_{11}\frac{\partial\mu_1}{\partial x} - M_{1e}\frac{\partial\phi}{\partial x} \tag{7-13}$$

$$J_e = -M_{e1}\frac{\partial\mu_1}{\partial x} - M_{ee}\frac{\partial\phi}{\partial x} \tag{7-14}$$

If $\partial\phi/\partial x = 0$, the effective charge for the solute (q) is given by the equation

$$q = \left(\frac{J_e}{J_1}\right)_{\nabla\phi=0} = \frac{M_{e1}}{M_{11}} \tag{7-15}$$

q is thus the ratio of the charge flux to the solute flux at zero electric potential gradient. Equivalently, it is the charge that must be removed from a small volume, per unit solute flux, if the small volume is to stay neutral. From the Onsager relations, $M_{1e} = M_{e1}$, so Eq. (7-13) can be rewritten

$$J_1 = -M_{11}(\nabla\mu_1 + q\,\nabla\phi) \tag{7-16}$$

This equation is completely analogous to the equation used in Chap. 5 for ionic solids. The primary difference between metals and ionic conductors is that in metals electrons carry the great majority of the current. However, from a phenomenological viewpoint, since a potential gradient gives rise to a matter flux, there must be an effective charge associated with the solute. The physical nature of this effective charge is for now immaterial, as is the observation that the transport number of the solute may be only 10^{-5}.

Physical Interpretation of q. If an interstitial solute migrates under the influence of a direct current, Eq. (7-16) can be used to determine q. In principle, a steady state can be set up and q determined from the steady-state concentration gradient. Alternatively, the solute flux across a plane can be measured, and if the diffusion coefficient is known, q can be calculated. Experiments of this latter type have been performed for carbon in austenite,[1] and it is found that q is positive and has a magnitude of 3.7 electron units. By analogy with ionic solids and liquid electrolytes, one can say that carbon in austenite migrates as if it had a valence of $+3.7$. This is also consistent with the fact that carbon has a valence of $+4$ in most of its compounds. Similar experiments show that $q \simeq 1$ for hydrogen in palladium[2] and $q \simeq 7$ for nitrogen in iron.[3] These also roughly agree with the valence of these interstitials in compounds. However, a critical examination of just what is meant by the valence of a solute in a metal leads to considerable doubt that there could be any simple connection between the normal valence of the solute in a compound and the force exerted by a current on the solute atom in solution in the metal.

As a first attempt to integrate these results into a modern theory of metals, one could say that q is the charge of the solute nucleus plus the electrons that are bound to it in the metal. This might indeed bear a relation to the valence of the ion in compounds. However, the question of how tightly bound an electron must be to be called "bound" quickly leads to a very difficult problem.

An even larger uncertainty enters the interpretation of q when the effect of the electron flux is considered. The flowing electrons are scattered by the atoms in the lattice and will be scattered much more strongly by an imperfection such as an interstitial or a diffusing atom at the saddle point in a jump. When electrons are scattered by, or "bounce off," such defects, they transfer momentum to the defect. This gives rise to what has been called an electron "breeze," which tends to "blow" the atoms in the direction that the electrons are moving. This electron breeze will tend to make q for an atom negative no matter what the net charge on the atom, since the breeze always blows toward the positive electrode.

The results given above for interstitial solute can best be rationalized if the electron breeze is ignored. However, another set of experiments indicates that the scattering of electrons may be quite important. Consider the case of a pure noble metal such as gold or copper. It is thought that the free-electron theory of metals can safely be applied here. When this is done, the resulting model consists of atoms with a net charge of $+1$ immersed in a gas of the free electrons. Diffusion occurs by a vacancy mechanism; and ignoring electron motion, one can talk of an atom with a $+1$ charge or vacancy with an effective -1 charge. This naïve model would indicate that $q = +1$ for the gold or copper atoms. However, Huntington and Grone have determined

[1] P. Dayal and L. Darken, *Trans. AIME*, **188**: 1156 (1950).
[2] C. Wagner and G. Heller, *Z. Physik Chem.* (Leipzig), **46B**: 242 (1940).
[3] J. C. M. Li, private communication.

q for pure gold and found that the atoms move toward the positive electrode and the experimental value of q decreases from -9 at 750°C to -6 at 1000°C.[1] Thus they conclude that the primary force on the atom at the saddle point stems from the electron breeze. The actual calculation of the magnitude of the momentum transferred from the electrons to the diffusing atom is quite difficult and has yet to be treated completely. The difficulties are pointed out, and an approximate calculation has been given by Huntington and Grone.[1] Similar experiments on pure copper lead to an even more pronounced variation of q with temperature. Two separate groups have found that q is negative below roughly 950°C and becomes positive above this temperature.[2] This marked temperature dependence is not understood.

An extensive series of studies of the electrolysis of solid alloys has shown that this effect is quite general. In view of the uncertainty of the interpretation of the results in pure metals and interstitial alloys, no attempt will be made to discuss the results found in substitutional alloys and intermetallic compounds. Most of this work has been done in Germany by Seith,[3] Heumann, and Wever. The interested reader is referred to this literature.

PROBLEMS

7-1. Consider the "thermocouple" shown in Fig. 7-3 with one end at temperature T_1 and the open end at T_2. An interstitial solute is introduced for which $Q^* = -20$ kcal/mole in one leg and $Q^* = 0$ in the other leg. Assuming that the solute

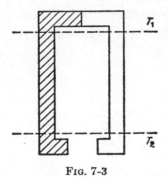

FIG. 7-3

[1] H. Huntington and A. Grone, *J. Phys. Chem. Solids*, **20**: 76 (1961).

[2] H. Wever, *Z. Elektrochem.*, **60**: 1170 (1956). A. Grone, *J. Phys. Chem. Solids*, **20**: 88 (1961).

[3] The work done prior to 1955 is summarized by W. Seith, "Diffusion in Metallen," pp. 254ff, Springer-Verlag, Berlin, 1955. For a good review in English, see W. Jost, "Diffusion," chap. 8, Academic Press, Inc., New York, 1952.

concentration is the same in both phases at the interface at temperature T_1, derive the equation for the steady-state concentration difference ΔC between the two phases at T_2 as a function of $T_1 - T_2$. Using this equation, calculate ΔC if $T_2 = 700°C$ and $T_1 = 800°C$.

The analogy between this system and a normal thermocouple is complete. For example, if solute flows through the junction at temperature T_1, the heat adsorbed or evolved per mole of solute would be just the difference in the molar heats of solution for the solute in the two phases. The analogous heat in a thermocouple is called the Peltier heat. It is not common to talk about Q^* for the electrons in a metal, but an equation for the voltage $\Delta\phi$ produced at the open end of the junction of a thermocouple can be derived which corresponds to the equation for ΔC derived above. The equation for $\Delta\phi$ can be derived in terms of Q^* for the electrons but is more commonly expressed in terms of S^*, the entropy of transport for the electrons where $Q^* \equiv TS^*$. See, for example, R. Heikes and R. Ure (eds.), "Thermoelectricity: Science and Engineering," chap. 1, Interscience Publishers, Inc., New York, 1961.

7-2. In a pure metal the equilibrium fraction of vacant sites at any temperature is given by the equation $N = \alpha \exp(-\Delta H_v/RT)$. Given a pure metal specimen in a temperature gradient, if the equilibrium vacancy concentration is maintained in each isothermal sheet, derive the equation relating the vacancy flux to ΔH_v, Q_v^*, and dT/dx.

7-3. Thermal diffusion arises from the fact that in a temperature gradient atoms jumping from one plane to an adjacent hotter plane do so with a slightly different frequency than that of atoms making the reverse jump. Estimate the ratio of these two jump frequencies using Eqs. (7-9) and (7-10) and taking $dT/dx = 500°C/cm$, $|H_0 - H_f| = 20$ kcal/mole, and $T = 1000°K$.

7-4. It was stated in the text that the transport number of carbon in γ-Fe is roughly 10^{-5} at $1200°C$, i.e., one part in 10^5 of the electric current is carried by carbon atoms at this temperature. Taking $Q_c^* = -23$ kcal/mol, $D = 10^{-7}$ cm^2/sec, and the thermal conductivity of ferrite to be 0.1 cal/°C sec cm, calculate the fraction of the heat carried by carbon atoms in ferrite when a homogeneous 0.01% C alloy is placed in a temperature gradient at $1000°K$.

INDEX

Activated complexes, 58
Activated state, 57
Activation energy, empirical rules, 65–66
  (*See also* ΔH_m)
Activation entropy (*see* ΔS_v)
Activation volume, definition of, 81
  determination of, 82
Anelasticity due to diffusion, interstitials in bcc, 87–89
  substitutional alloys, 94

Brownian motion, 40

Capture radius, 28
Chemical diffusion coefficient, 115n.
  relation to self-diffusion coefficient, 125
  variation with composition, 133–134
Compound semiconductors, 155–160
Correlation effects, in pure metals, 100
  in very dilute alloys, 106–109
Correlation factor, calculation of, 102
  definition of, 101

D_0, Zener's theory of, 62–65
ΔH_m, determination of, in ionic solids, 147
  in metals, 77–78
  effect on, of atomic size, 99
  of valence, 98
  theoretical calculation of, 70

ΔH_v, definition of, 56
  determination of, 71–74, 76
  effect on, of atomic size, 99
  of valence, 97
  theoretical calculation of, 68–70
ΔS_v, definition of, 56
  determination of, 71–74
  theoretical calculation of, 67
Diffusion coefficient, anisotropy, 34–36
  chemical (*see* Chemical diffusion coefficient)
  definition of, 2
  determination, concentration dependence, 30
    elastic aftereffect, 90
    resonance techniques, 90
    steady-state, 3
    thin-film technique, 9
  effect on, of crystal structure, 134
  of impurities, 145–151
  of pressure, 81–83
  at grain boundaries, 167–175
  from random-walk analysis, 51–54, 84
  relation to ionic conductivity, experiment, 143–151, 154–155
  theory, 141–143
  temperature dependence, 61
  variation with composition, concentrated alloys, 133
  dilute alloys, 112
Diffusion equations (*see* Fick's first law; Fick's second law)
Dislocations, contribution, to apparent D_l, solute D_l, 178
  to solvent D_l, 175–178
  D at, 176–177

Divacancies, definition of, 78
    equilibrium concentration, 78–79
Divacancy, migration into, 80

Elastic aftereffect, 90
Electrolysis of solids, 196
Electron hole, 158–160
Elemental semiconductors, 160–161
Enthalpy of activation (*see* ΔH_m)
Entropy of activation (*see* ΔS_s)
Error function, definition of, 12

Fick's first law, definition of, 2
    and random movement, 42
Fick's second law, definition of, 5
    solutions, D function, of composi-
        tion, 29
        of time, 31
    finite systems, 15, 16
    infinite systems, 7–15
    steady-state, 2
Field emission, 186
Frenkel disorder, definition of, 139
    effect of impurities on conductivity,
        149–151
    equilibrium concentration, 139
    relation of conductivity to D_T, 151–
        156

Grain boundary diffusion, analysis of,
    tracer technique, 166–169
    anisotropy in tilt boundaries, 174
    contribution to apparent D_l, 170
    effect on, of boundary misorienta-
        tion, 172
        of temperature, 172
Grain boundary diffusion coefficient,
    definition of, 167
    determination of, 169–170
Grain boundary grooving, analysis of,
    179–184
    contributions of volume, surface,
        and gaseous diffusion, 181–182
    determination of D_s, 184–186

Heat of transport, 189
    kinetic model, 191–196
    phenomenological definition of, 191

Impurity self-diffusion, 95
    data, in Ag and Cu, 111
    size effects, 98
    valence effects, 96
Intermetallic compounds, diffusion in,
    162
Internal friction (*see* Anelasticity)
Interstitial, definition of, 46
Interstitialcy mechanism, 151–154
Ionic conductivity, effect of impuri-
    ties on, 145–151
    extrinsic range of, 147
    intrinsic range of, 146
    relation to D, experimental, 143–
        151, 154–155
    theory, 141–143
Ionic solids, point defects in, 137–139
    (*See also* Frenkel disorder; Schottky
        disorder)

Jump frequency, calculation of, 57–61
    relation to D, 52, 53

Kirkendall effect, 116
    Darken's analysis, 117

Marker movement (Kirkendall ef-
    fect), 116, 117
Mechanisms of diffusion, 43–47
    interstitial, 43
    interstitialcy and Crowdion, 45
    vacancy, 44
Mobility, 24, 140–141n.
    relation to D in alloys, 124
Multiphase binary systems, 132

Noncubic lattices, diffusion in, 32–36

Ordered alloys, 162

Permeability, 4
Phenomenological equations, 122
  electrolysis, 197
  thermal diffusion, 190
Point defects (*see* type of defect, e.g.,
    Interstitial; Vacancies concen-
    tration; etc.)
Precipitation, kinetics of, 19–23
Pressure effect on diffusion, 81–83

Quenching experiments, 75–76

Random walk, analysis of, 47–51
  equation for D, 51–54, 84
Relaxation time, D for interstitials, 89
  definition of, 18
  excess vacancies, 76–77
  magnetic measurement of, 93, 94
  resonance, 91, 92

Saddle point, 59
Schottky disorder, definition of, 139
  effect of divalent impurities, 145
  equilibrium concentration, 139

Self-diffusion coefficient, definition of,
    95n.
  (*See also* Impurity self-diffusion)
Semiconductors, compound, 155–160
  elemental, 160–161
Soret effect (*see* Thermal diffusion)
Stress-assisted diffusion, 23–28
Surface diffusion coefficient, defini-
    tion of, 182
  determination of, 184

Ternary alloys, 130
Thermal diffusion, definition of, 189
  relation to thermocouple, 199

Vacancies concentration, in diffusion
    couple, 127
  at equilibrium, 54–56
  definition of, 46
  in ionic solids, 138, 139
Vacancy formation, energy of (*see*
    ΔH_v)
  mechanism of, 128

Zener's theory of D_0, interstitials, 62–
    64
  vacancy mechanism, 64–65